Springer Transactions in Civil and Environmental Engineering

Editor-in-Chief

T. G. Sitharam, Indian Institute of Technology Guwahati, Guwahati, Assam, India

Springer Transactions in Civil and Environmental Engineering (STICEE) publishes the latest developments in Civil and Environmental Engineering. The intent is to cover all the main branches of Civil and Environmental Engineering, both theoretical and applied, including, but not limited to: Structural Mechanics, Steel Structures, Concrete Structures, Reinforced Cement Concrete, Civil Engineering Materials, Soil Mechanics, Ground Improvement, Geotechnical Engineering, Foundation Engineering, Earthquake Engineering, Structural Health and Monitoring, Water Resources Engineering, Engineering Hydrology, Solid Waste Engineering, Environmental Engineering, Wastewater Management, Transportation Engineering, Sustainable Civil Infrastructure, Fluid Mechanics, Pavement Engineering, Soil Dynamics, Rock Mechanics, Timber Engineering, Hazardous Waste Disposal Instrumentation and Monitoring, Construction Management, Civil Engineering Construction, Surveying and GIS Strength of Materials (Mechanics of Materials), Environmental Geotechnics, Concrete Engineering, Timber Structures.

Within the scopes of the series are monographs, professional books, graduate and undergraduate textbooks, edited volumes and handbooks devoted to the above subject areas.

The proposal for each volume is reviewed by the main editor and/or the advisory board. The chapters in each volume are individually reviewed single blind by expert reviewers (at least four reviews per chapter) and the main editor.

Sarat Kumar Das · Krishna R. Reddy ·
Lohitkumar Nainegali · Surabhi Jain
Editors

Geoenvironmental and Geotechnical Issues of Coal Mine Overburden and Mine Tailings

 Springer

Editors
Sarat Kumar Das
Department of Civil Engineering
Indian Institute of Technology (ISM)
Dhanbad
Jharkhand, India

Krishna R. Reddy
Department of Civil, Materials,
and Environmental Engineering
University of Illinois Chicago
Chicago, IL, USA

Lohitkumar Nainegali
Department of Civil Engineering
Indian Institute of Technology (ISM)
Dhanbad
Jharkhand, India

Surabhi Jain
Department of Civil Engineering
Indian Institute of Technology (ISM)
Dhanbad
Jharkhand, India

ISSN 2363-7633   ISSN 2363-7641 (electronic)
Springer Transactions in Civil and Environmental Engineering
ISBN 978-981-99-6293-8   ISBN 978-981-99-6294-5 (eBook)
https://doi.org/10.1007/978-981-99-6294-5

This Springer imprint is published by the registered company Springer Nature Singapore Pte Ltd.
The registered company address is: 152 Beach Road, #21-01/04 Gateway East, Singapore 189721,
Singapore

Paper in this product is recyclable.

Preface

The consequence of massive coal mining is the generation of mine overburden of quantity double the amount of coal mined. The requirement of usable or forest lands for storage, dust and plume gas from stockpiles, acid mine drainage, occasional landslides of dump slopes, and environmental hazard in terms of air, water and land contamination, are a few problems associated with mine overburden. Large heterogeneity in gradations and complex mineralogical compositions are major impending factors for its utilization. Another consequence of mining metals and minerals is the dams that are built to store tailings, the waste leftover from mining. Over the last ten years, 320 tailing dams have been constructed, more than 1700 tailing dams exist all over the world, and 687 dams are classified as high risk. Failures of such tailing dams have resulted in incomparable environmental and infrastructure disruption, as well as human casualties.

This book is inspired by the idea of educating about significant geoenvironmental and geotechnical issues, practical challenges encountered, and solutions adopted with a focus majorly on coal mine overburden and mine tailings. Further, the book aims to provide knowledge-based information for diverse readers (researchers, practitioners, and educators) to assess, monitor, and manage coal mine overburden and mine tailings. Several authors that are well-known specialists, experts from industry and academia in distinct fields, have contributed the chapters covering a wide range of topics. The book covers topics such as mine overburden and tailing management, mine backfilling and stabilising via various processes, CPTu field-based soil behaviour indices, landfill liners and barrier systems, geochemical, microbial and environmental aspects of acid mine drainage, treatment techniques, and mineral carbonation of mine tailings. The usage of various tailings such as bauxite residue and fly ash in the construction, and emerging technology of carbon capture and storage of mine tailings is also discussed.

Engineers and researchers will benefit from it as they improve and advance their methodologies. Academicians, researchers, working professionals, and notably students in the geoenvironmental and geotechnical fraternity will find this book to be a valuable resource.

Jharkhand, India	Sarat Kumar Das
Chicago, USA	Krishna R. Reddy
Jharkhand, India	Lohitkumar Nainegali
Jharkhand, India	Surabhi Jain

Contents

1   **Geotechnical Considerations of Mine Tailings Management through Mine Backfilling** 1
Seneth Jayakodi, Nagaratnam Sivakugan, and Peter To

2   **CPTu-Based Soil Behaviour Type Indexes that are Independent of Sleeve Friction Readings: An Application in Tailings** 31
Luis Alberto Torres-Cruz, Nico Vermeulen, and Abideen Owolabi

3   **Assessment of Mine Overburden Dump Stability Using Numerical Modelling** 39
Tarun Kumar Rajak and Laxmikant Yadu

4   **Performance of Coal Mine Overburden Dump Slope Under Earthquakes Using Extended Finite Element Method Based Voronoi Tessellation Scheme** 63
Madhumita Mohanty, Rajib Sarkar, and Sarat Kumar Das

5   **Slope Stability Analysis of Coalmine Overburden Dump Using a Probabilistic Approach** 75
Ashutosh Kumar, Sarat Kumar Das, Lohitkumar Nainegali, and Krishna R. Reddy

6   **Suitability of Bauxite Residue as a Landfill Liner Material—An Overview** 89
Narala Gangadhara Reddy, Tayyaba Siddiqua, Manikanta Devarangadi, and Chandra Bogireddy

7   **Fly Ash Based Geopolymer Modified Bitumen (GMB) binder—An Overview** 101
Bojjam Sravanthi and N. Prabhanjan

**8   Mineral Carbonation of Mine Tailings for Long-Term Carbon
       Capture and Storage** 109
       Faradiella Mohd Kusin and Verma Loretta M. Molahid

**9   Environmental Sustainability Assessment of Alternative
       Controlled Low Strength Materials as a Fill Material** 133
       Anshumali Mishra, Sarat Kumar Das, and Krishna R. Reddy

**10   State of the Art Review on the Geochemical, Microbial
       and Environmental Aspects of Passive Acid Mine Drainage
       Treatment Techniques** 147
       M. K. Kaushik

**11   Acid Mine Drainage and Metal Leaching Potential at Makum
       Coalfield, Northeastern India** 177
       Sk. Md. Equeenuddin, S. Tripathy, Prafulla Kumar Sahoo,
       and M. K. Panigrahi

Editors and Contributors

About the Editors

Dr. Sarat Kumar Das is a professor and head of the Department of Civil Engineering at the Indian Institute of Technology, Indian School of Mines, Dhanbad, India. Professor Das obtained his bachelor's in Civil Engineering from the College of Engineering and Technology (presently OUTR) Bhubaneswar, India, followed by his master's and Ph.D. in Geotechnical Engineering from Indian Institute of Technology Kanpur (IITK), India. Supported by the Endeavor Research Fellowship of the Australian Government, he undertook his postdoctoral training at James Cook University, Townsville, Australia. Professor Das has about 30 years of experience in the field of civil and geotechnical engineering in academia and industry. His research interests include geoenvironmental engineering, AI and optimization methods, and biogeotechnics.

Dr. Krishna R. Reddy is a professor of Civil and Environmental Engineering, director of Sustainable Engineering Research Laboratory, and director of the Geotechnical and Geoenvironmental Engineering Laboratory at the University of Illinois at Chicago, USA. He received his Ph.D. in Civil Engineering from the Illinois Institute of Technology, Chicago. He received gold medals during B.E. in Civil Engineering at Osmania University and M.E. in Civil Engineering at the Indian Institute of Technology, Roorkee. His teaching and research expertise includes civil and geotechnical infrastructure systems, water and environmental pollution remediation, waste management and recycling, and sustainable and resilient engineering. Dr. Reddy has received several awards, including ASCE Wesley W. Horner Award, ASTM Hogentogler Award, the UIC Distinguished Researcher Award, University of Illinois Scholar Award, and University of Illinois Award for Excellence in Teaching.

Dr. Lohitkumar Nainegali is an assistant professor at the Department of Civil Engineering, Indian Institute of Technology (ISM) Dhanbad. He received his M.Tech. and Ph.D. in Civil Engineering from the Indian Institute of Technology Kanpur, India.

His research is mainly focused on the physical and numerical modeling of geotechnical systems, stability of geotechnical structures, soils dynamics and foundations, and valorization of industrial waste/utilization of mine overburden. He has received various research grants on similar research areas from the Department of Science and Technology, India and Coal India Limited, etc. and has conducted several consultancy projects. He has several publications in renowned international journals and conferences. He has mentored one Ph.D. and 07 M.Tech. scholars and is currently guiding four Ph.D. scholars.

Dr. Surabhi Jain is currently pursuing postdoctoral research at the Department of Civil Engineering, Indian Institute of Technology (ISM) Dhanbad. She was awarded the PECFAR award in 2022 and working in the collaboration with Technical University of Munich, Germany. She has completed one year of post-doctorate at Guangdong Technion-Israel Institute of Technology, Shantou, China. She received her Ph.D. from Indian Institute of Technology Madras, Chennai, India. Her research work is in biogeotechnics, utilization of industrial waste in construction, and sustainable engineering. She is the first researcher who started working on valorizing industrial waste using biochemical processes. She has delivered various lectures on biogeotechnics and biocementation on different reputed platforms. Please make sure that the below mentioned BCC should be updated in cover. This contributed book edited by leading global experts focuses on the geoenvironmental and geotechnical issues of coal mine overburden and mine tailings and its unengineered dumping. It aims to provide knowledge-based information for diverse readers to assess, monitor, and manage coal mine overburden and mine tailings in various engineering applications while highlighting efficient solutions to reutilize the waste and conserve natural resources leading to sustainable development. The content also assesses mine backfilling, techniques to stabilize mine tailing storage facilities, mineral carbonation of mine tailings, landfill liners and barrier systems, reclamation of coal mine overburden, and geochemical, microbial, and environmental sustainability assessment, among others. This book is a useful resource for those in academia and industry.

Contributors

Chandra Bogireddy Shantou University, Shantou, Guangdong, China

Sarat Kumar Das Indian Institute of Technology (Indian School of Mines), Dhanbad, Jharkhand, India

Manikanta Devarangadi Ballari Institute of Technology and Management, Ballari, India

Sk. Md. Equeenuddin National Institute of Technology, Rourkela, Odisha, India

Seneth Jayakodi College of Science and Engineering, James Cook University, Townsville, Queensland, Australia

M. K. Kaushik DAV Institute of Engineering and Technology, Jalandhar, Punjab, India

Ashutosh Kumar Indian Institute of Technology (Indian School of Mines), Dhanbad, Jharkhand, India

Faradiella Mohd Kusin Department of Environment, Faculty of Forestry and Environment, Universiti Putra Malaysia, UPM, Serdang, Selangor, Malaysia; Institute of Tropical Forestry and Forest Products (INTROP), Universiti Putra Malaysia, UPM, Serdang, Selangor, Malaysia

Anshumali Mishra Indian Institute of Technology (Indian School of Mines), Dhanbad, Jharkhand, India

Madhumita Mohanty Indian Institute of Technology (Indian School of Mines), Dhanbad, Jharkhand, India

Verma Loretta M. Molahid Department of Environment, Faculty of Forestry and Environment, Universiti Putra Malaysia, UPM, Serdang, Selangor, Malaysia

Lohitkumar Nainegali Indian Institute of Technology (Indian School of Mines), Dhanbad, Jharkhand, India

Abideen Owolabi University of the Witwatersrand, Johannesburg, South Africa

M. K. Panigrahi Indian Institute of Technology, Kharagpur, West Bengal, India

N. Prabhanjan Department of Civil Engineering, SR Engineering College, Warangal, India

Tarun Kumar Rajak Department of Civil Engineering, Shri Shankaracharya Institute of Professional Management and Technology, Raipur, India

Krishna R. Reddy University of Illinois Chicago, Chicago, IL, US

Narala Gangadhara Reddy Kakatiya Institute of Technology and Science, Warangal, India; School of Building and Civil Engineering, Fiji National University, Samabula, Suva, Fiji

Prafulla Kumar Sahoo Central University of Punjab, Bathinda, Punjab, India

Rajib Sarkar Indian Institute of Technology (Indian School of Mines), Dhanbad, Jharkhand, India

Tayyaba Siddiqua Kakatiya Institute of Technology and Science, Warangal, India

Nagaratnam Sivakugan College of Science and Engineering, James Cook University, Townsville, Queensland, Australia

Bojjam Sravanthi Kakatiya Institute of Technology and Science, Warangal, India

S. Tripathy Indian Institute of Technology, Kharagpur, West Bengal, India

Peter To College of Science and Engineering, James Cook University, Townsville, Queensland, Australia

Luis Alberto Torres-Cruz University of the Witwatersrand, Johannesburg, South Africa

Laxmikant Yadu Department of Civil Engineering, National Institute of Technology, Raipur, India

Nico Vermeulen Jones and Wagener (PTY) LTD, Johannesburg, South Africa

Chapter 1
Geotechnical Considerations of Mine Tailings Management through Mine Backfilling

Seneth Jayakodiⓘ**, Nagaratnam Sivakugan**ⓘ**, and Peter To**ⓘ

1.1 Mining and Mine Backfilling

Mining is a complex but systematic process of extracting valuable minerals from ore deposits. Currently, almost every continent in the world except Antarctica is undergoing mining activities extensively and the major contributing countries include Australia, Brazil, Canada, China, India, Poland, South Africa and USA. Nature has bestowed Australia generously with valuable minerals and the amount the country has mined out thus far is considerably less compared to the amount it reserves. Main Australian mining commodities are Gold, Zinc, Copper, Lead, Coal, Silver, Diamonds, Mineral Sands and Bauxite for which Australia is known as the world's largest producer. Mining contributes a substantial proportion to the country's Gross Domestic Product (GDP) and is the backbone of the Australian economy. Mining of all resources contributes about 10% to Australia's GDP and around 60% of total export revenue [1]. The BHP Group, Rio Tinto, Glencore, Fortescue Metals, Newcrest Mining, South32, Northern Star Resources and Evolution Mining are some major mining companies operating in Australia. Figure 1.1 shows the locations of major mines and mineral deposits in Australia.

Mining can be categorised as open pit mining and underground mining based on the orientation and location of the deposit. When the ore deposit is at shallow depth and spreading more longitudinally, it can be mined out by removing the overburden as the open pit mining method. In contrast, when the ore body is located deep inside the ground, underground mining techniques are adopted to reach the ore body through shafts and cross-cut tunnels, and to extract the valuable ore where the removal of ore generates large voids, known as stopes. These stopes are accessed through horizontal drives located along the ore body, allowing access to machinery

S. Jayakodi (✉) · N. Sivakugan · P. To
College of Science and Engineering, James Cook University, Townsville, Queensland, Australia
e-mail: seneth.jayakodi@my.jcu.edu.au

© The Author(s), under exclusive license to Springer Nature Singapore Pte Ltd. 2024   1
S. K. Das et al. (eds.), *Geoenvironmental and Geotechnical Issues of Coal Mine Overburden and Mine Tailings*, Springer Transactions in Civil and Environmental Engineering, https://doi.org/10.1007/978-981-99-6294-5_1

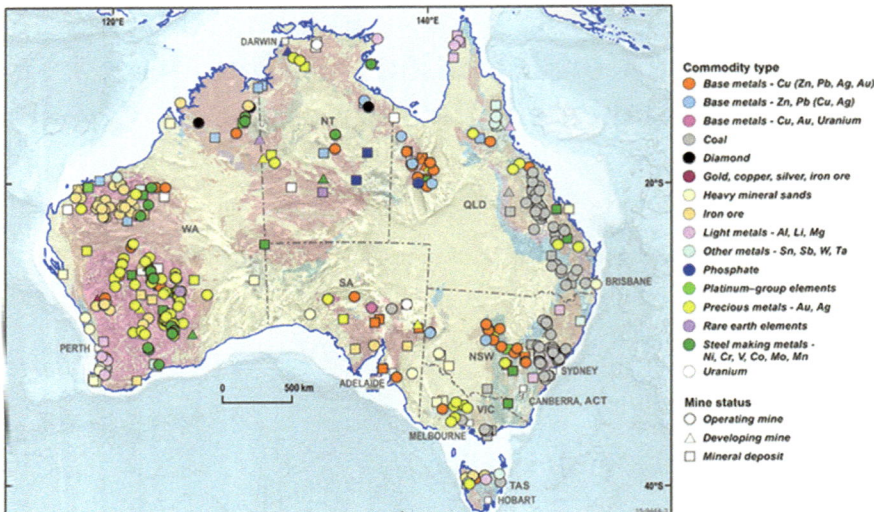

Fig. 1.1 Major mining and mineral deposits in Australia, 2016 (*Source* Major mining and mineral deposits in Australia, 2016 https://www.ga.gov.au/education/classroomresources/minerals-energy/australian-mineral-facts by Geoscience Australia which is © Commonwealth of Australia and is provided under a Creative Commons Attribution 4.0 International Licence and is subject to the disclaimer of warranties in section 5 of that licence)

and other underground services. Underground mine stopes have base dimensions of 10–15 m or more and heights as high as 100 m or more, and they are assumed rectangular in cross section for analytical purposes [2]. Once the ore has been removed and processed, the stope requires backfill for surrounding stability and continued production. The backfilling of stopes can represent up to 20% of underground mining costs [3]. The horizontal drives need to be barricaded or capped to allow for the backfill process to take place. The construction and performance of these barricades is a major topic in the mining industry today. The correct calculation of the stress state in the backfill and the associated pressures exerted on the barricade is a critical engineering challenge [4].

Mine backfilling is an essential step in the mining cycle for artificially supported underground mines [5]. An efficient backfill system provides several operational and economic benefits such as maximum ore extraction, minimum ore dilution, improvement of regional stability around the ore body, avoid the risk of land subsidence of mined out premises, ensure the safety of miners, etc. In addition, backfilling plays a vital role in the disposal of mine tailings with a minimum impact to the environment, thus ensuring sustainable mining practice. The mining industry in Australia is also the largest generator of solid wastes [6], making the backfilling of voids and the disposal of tailings an integral part of the mining process [7]. The mine tailings are waste material generated during the mining process which should be carefully disposed to Tailing Storage Facilities (TSF) surrounded by tailing dams [8], and/

or used to backfill the underground voids. The underground voids created annually in Australia can reach 10 million cubic metres [3]. Though the backfill system is often expensive, typically covering about 30% of total mining costs, reliability and flexibility of the backfill system greatly support the mining operation [9]. Backfilling gives an extra advantage over tailing dams as it rules out the cost for constructing tailing dams, negates the acquisition of massive land surface for TSFs, and reduces the negative environmental impacts.

1.2 Types of Mine Backfills

Backfills are generally processed mine tailings and are in the form of Hydraulic fills, Paste fills, Aggregate fills, Rock fills, Sand fills, etc. The choice of materials that are utilized for backfill operation depends on the location of backfilling, availability of materials for a particular fill, physicochemical properties of the materials, and prevailing ground conditions. Backfills are of two types according to cohesiveness namely Granular backfills (no cohesion) and Cohesive backfills. Further, backfills fall into two main categories based on the backfilling strategy being used as Cemented and Uncemented.

1.2.1 Uncemented Backfills

Uncemented backfills have no binding agents mixed thus mechanical behaviour and performance can be studied using soil mechanics theories. This includes Sand fills (SF), Rock fills (RF), Aggregate fills (AF), and Hydraulic fills (HF).

Hydraulic Fills (HF)

Hydraulic fills are sandy silts (ML) or silty sands (SM) with no clay fraction where the fine fraction is removed by the desliming process using hydrocyclones. Physicochemical properties of the fill material are among the first few important aspects, to understand the suitability of the material for a particular backfilling process. Particle size distribution, specific gravity, bulk density, Atterberg limits, relative density, permeability, angularity of grains, friction angle, etc. are the main physical characteristics of the fill material while the chemical properties include the presence of mineral phases, elemental composition, acidity/alkalinity, total organic compounds, heavy metal concentration, etc. [10, 11]. According to Rankine et al. [12] and Sivakugan et al. [13], based on the analysis of more than 25 HFs from different major mines in Australia, it has been found that all fills fall within a narrow particle size distribution band and the specific gravity of the grains are in the range of 2.8–4.5 due to presence of heavy metals. HFs have very sharp, angular grains, giving higher friction angles (ϕ) than those for natural soils. All HF slurries settling only under self-weight, manage to settle at high relative densities (D_r) of 40–70%. According to Pettibone and Kealy [14], the in-situ measurements showed D_r values ranging from 44 to 66% at four different mines in the United States and that confirms the above laboratory test results by Rankine et al. [12] and Sivakugan et al. [13]. Further, HF has void ratio (e) of 0.67, and porosities (n) of 37–49% and dry unit weight (γ_d) in kN/m^3 of 5.7 times specific gravity $[\gamma_d(\text{kN/m}^3) = 5.7G_s]$. Typical γ_d of HF is in the range of 16–25 kN/m^3 [12]. The lower cost for production and transportation through pipelines, and the simplicity in handling are the main advantages of hydraulic fills.

Behaviour of Hydraulic Fill during Backfilling Operation

Hydraulic fills are initially transported to the stope in the form of a slurry through pipelines, at solid contents of 65–75% [12, 13]. The horizontal drives blocked by a barricade wall, made of special porous bricks with permeability (k) 2 or 3 orders of magnitude greater than that of HF [12], retains the fill and allows the water to drain through. The horizontal access drives are often located at more than one level. Initially, the drives located at upper levels of the vertical stope act as exit points for the decanted water, and also serve as drains when the HF rises in the stope [13]. Filling the stope does not occur instantaneously or continuously at one stretch instead it is carried out as layer filling with breaks/laps at certain filling intervals, allowing for drainage (e.g., 12 h filling and 12 h resting).

Drainage Considerations

One of the primary causes of barricade failures is liquefaction which is often referred as 'mud rush' by miners. This scenario occurs when the pore water pressure builds up and reduces the effective stresses and hence the shear strength within the stope as a result of poor drainage in the hydraulic fill system.

Herget and De Korompay [15] have suggested that the minimum hydraulic conductivity/permeability (k) of HF should be 100 mm/h in order for the fill to perform satisfactorily. When the permeability is high, faster removal of water from the stope helps to increase the stability of the mine fill stope. Grice [16] has suggested that the effective grain size $D_{10} > 10\ \mu$m will ensure adequate drainage.

Sivakugan et al. [13] observed that when $k = 7$–35 mm/h, HF systems in the mine stopes performed well, suggesting that Herget and De Korompay [15] threshold value is too conservative. D_{10} for adequate drainage was found to be 10–40 μm, satisfying Grice's [16] recommendation. However, the anecdotal evidence and back calculations using the measured flow in the mine stopes suggest that the permeability of the HF in the mine is often larger than what is measured in the laboratory under controlled conditions [7]. Brady and Brown [17] and Kuganathan [18] proposed permeability values in the range of 30–50 mm/h, which are significantly larger than those measured in the laboratory for similar fills but still their values are much less than the threshold limit prescribed by Herget and De Korompay [15], which is 100 mm/h. Figure 1.2 shows drainage pipes fixed to the reinforced concrete barricade wall to release water out from the HF slurry and to avoid building up the pore water pressure within the fill.

Fig. 1.2 Drainage pipes fixed to the reinforced concrete barricade to drain the water from hydraulic fill slurry

Stability Considerations

The stability of the HF stope depends on several parameters that govern the strength and stiffness of the fill which are directly related to the relative density of the fill. When the HF is denser, the relative density (D_r) and friction angle (ϕ) are higher, and thus the fill is more stable. Oedometer tests on HF showed significant creep settlements that took place on the completion of consolidation settlements [13].

1.2.2 Cemented Backfills

Cemented backfills include a small amount of binding agent such as cement, fly ash, gypsum, blast furnace slag to improve strength. Following fill types are some examples for cemented backfills.

Cemented hydraulic fills (CHF) have grain size < 420 μm and are more similar to paste fill but the significant difference is larger particle size distribution (PSD) compared to paste fill.

Cemented rock fill (CRF) is a mixture of RF with CHF (RF:CHF = 1:1 to 3:1 by weight) and the properties vary within the stope as the two fills segregate during placement.

Cemented aggregate fill (CAF) is a mixture of AF with CHF (AF:CHF = 1:3 by weight) and it suffers from segregation thus properties vary within the stope like in CRF.

Paste Fills (PF)

Paste fills use a very fine fraction of the tailings. A rule of thumb is that 15% of particles should be less than 20 μm, with a typical effective particle size (D_{10}) of 5 μm [19]. PF is a mixture of tailing material with a small percentage of binder, in the order of 3–6% by weight, and water [13]. The PSD is finer than HF and CHF, but have negligible colloidal fraction finer than 2 μm, and the maximum particle size is around 350–400 μm [13]. However, during transport, Pullum [19] has shown stratification of paste during pipe flow with all paste fills with a maximum particle size of over 20 μm. Paste fills with the maximum particle size $D_{max} < 20$ μm tend to form homogeneous paste fills during both transportation and deposition.

Paste fills fall into the broad category of thickened tailings which was introduced by Dr Ely Robinsky in the mid 1970s [20]. It is the densest form of backfill in the spectrum of thickened tailings placed underground as a backfill material [20, 21]. Thickened tailings are a special case of slurry tailings and tend to show many similar characteristics to paste, but paste fills are not same as thickened tailings. The primary difference is that the thickened tailings will segregate or settle out once a minimum velocity is reached. Hydraulic fills fall into the thickened tailings profile. However, a significant difference with the PF in terms of drainage is that the water content in paste fill is retained on placement, through the large surface area of the particles, eliminating the need for the design of drainage of the fill or barricades. Hence, the static and dynamic stability requirements should only be considered when designing paste filled stope systems. The static stability requirement is addressed by designing the PF system with an adequate strength to ensure that the vertical walls of the backfilled stopes remain stable throughout the mining of neighbouring stopes. If the paste becomes unstable, the adjacent walls may relax and displace into the open stope, causing high level of dilution and loss of mining economics. The dynamic stability of the PF system is achieved by designing the backfill mass to resist liquefaction or other seismic activities. There is a high risk of liquefaction in a PF system due to increased residual moisture content in the PF. According to Clough et al. [22], cemented sand with an Unconfined Compressive Strength (UCS) of 100 kPa was capable of resisting a seismic activity measuring 7.5 on the Richter scale and this value has been used by the mining industry as the minimum design strength of fill for any fill mass. The strength of PF satisfying static stability requirements is generally in excess of dynamic strength requirements. Because of no drainage requirement in PF system, the barricades are designed as temporary structures in PF stopes. Nevertheless, barricades must be designed with sufficient strength to retain the liquid mass of fill, until such time as it has cured properly to act as a plug at the base of the stope, thus preventing the additional deposited paste from entering the mine workings.

1.3 Stress Developments within Backfilled Mine Stopes

1.3.1 General

Mine stopes which have been generated by the ore removal during mining, are backfilled with processed mine tailings called backfills. A stope can be approximated as a cuboid with the cross-section dimensions of 10–15 m and vertical wall heights exceeding 100 m and thus the aspect ratio falls between 6 and 10. Backfilled or non-backfilled stopes are interconnected by horizontal access drives with a typical cross section of 5 m × 5 m or 6 m × 6 m, located at different sublevels of the stope, which allow access to mine machinery and other underground services.

1.3.2 Theory of Soil Arching

Arching phenomenon occurs when the frictional material moves against stable strata where the relative movement generates shear stresses along the interface that tend to hold the frictional material at its initial position. As result, the vertical stresses within the yielding material will be lower than the actual overburden stress, and that is known as arching [23].

Arching plays a significant role in many geotechnical and mining applications such as earth pressure on retaining walls [24–26], underground situations such as backfilled trenches and underground mine stopes [23, 27–32], pressures on piles and piled embankments [33–35], loadings on ditch conduits and pressures on buried structures [25, 36–40], designing storage silos (e.g., vessels storing granular materials such as chemical powders, capsules in pharmaceutical industry, flour, cement) [41, 42].

Arching Effect in Backfilled Mine Stopes

During filling, the vertical shear stresses acting on the stope walls can be significant. As a result, the vertical normal stress anywhere within the fill can be significantly less than the actual overburden pressure (γz) where γ is the unit weight of the fill and z is the fill height from top. Therefore, a substantial amount of the fill load is carried by the rock walls in the form of shear stresses [23].

1.3.3   Analytical Methods to Investigate Arching Effect

Some of the analytical solutions for arching and stresses within backfills include the 3D sliding wedge failure method [43], Simple arching theory and its modifications [29, 36, 40, 44], etc.

Mitchell et al. [43]:

In reality, backfilled stopes are surrounded by adjacent rock mass thus the backfill material is subjected to lateral confinement. The confined block mechanism (3D Sliding Wedge Failure) explained how the rock walls' support to reduce the fill stress due to arching effect in a confined environment like mine stope. The design strength required for stability is given by

$$UCS = \left(\gamma - 2\frac{c}{l}\right)\left[h - \frac{w}{2}\tan\alpha\right](\sin\alpha)(F) \qquad (1.1)$$

γ Bulk unit weight of the fill (kN/m^3)
c Cement bond strength of the fill/Cohesion (kPa)
l Length of the block (m)
h Height of the block (m)
w Width of the block (m)
α Angle of failure plane from horizontal ($= 45° + \phi/2$)
ϕ Friction angle of the fill
F Safety factor

   In the long term, the UCS of the fill is mainly due to binding agents, and strength contributed by friction can be neglected (i.e., $\phi = 0$). For frictionless material, $c = UCS/2$ ($q_u = 2c_u$). Then Eq. (1.1) becomes

$$UCS = \left(\gamma - \frac{UCS}{l}\right)\left(h - \frac{w}{2}\right)\frac{F}{\sqrt{2}} \qquad (1.2)$$

Equation (1.2) can be simplified further as below by assuming $F = \sqrt{2}$ and $h \gg l$

$$UCS = \frac{\gamma h}{h/l + 1} \qquad (1.3)$$

Marston's theory [36]:

Janssen [41] observed a significant vertical stress reduction in a corn-filled silo, compared to overburden stress (γz). Attributing this reduction to arching, an expression was developed using the limit equilibrium method to determine the average vertical stress at any given depth of the fill (z). This was later modified by Marston considering a 2D plane strain theory (for a trench) on arching and equations were developed to compute the vertical (σ_v) and horizontal (σ_h) normal stresses within cohesionless $(c = 0)$ mine fill stope as below.

$$\sigma_v = \frac{\gamma w}{2\mu K_a}\left[1 - exp\left(-\frac{2K_a\mu z}{w}\right)\right] \tag{1.4}$$

$$\sigma_h = K_a\sigma_v \tag{1.5}$$

$$K_a = \tan^2(45° - \phi/2) \tag{1.6}$$

γ Unit weight of the fill
w Stope width
z Fill depth from top of the fill
ϕ Friction angle of backfill
μ Coefficient of friction of backfill and wall (rock) $(\mu = \tan \delta)$
δ Interfacial friction angle between the wall and the fill (between $1/3\ \phi$ and $2/3\ \phi$)
K_a Rankine's active earth pressure coefficient

Terzaghi's theory [40]:

This includes the effect of cohesion into the Marston's equation enabling it to be used for any soil.

$$\sigma_v = \frac{(\gamma w - 2c)}{2K\ \tan\ \phi}\left[1 - exp\left(-\frac{2Kz\ \tan\ \phi}{w}\right)\right] \tag{1.7}$$

$$\sigma_h = K\sigma_v \tag{1.8}$$

$$K = \frac{1}{1 + 2\tan^2\ \phi} \tag{1.9}$$

c Cohesion of the fill
$\tan\ \phi$ Coefficient of internal friction of fill (same as μ, but with $\delta = \phi$)
K Lateral earth pressure coefficient

Aubertin et al. [44]: Modified Marston's theory

Marston's theory uses the active earth pressure coefficient (K_a) in the equation. However, the Aubertin et al. [44] solution (Eq. 1.10) is based on status of the stope wall during filling operation assuming the fill to be in active, passive or at-rest state.

$$\sigma_v = \frac{\gamma w}{2K\ \tan\ \phi}\left[1 - exp\left(-\frac{2Kz\tan\ \phi}{w}\right)\right] \tag{1.10}$$

$$\sigma_h = K\sigma_v \tag{1.11}$$

Active Earth Pressure Coefficient $K_a = \tan^2(45° - \phi/2)$

Passive Earth Pressure Coefficient $K_p = \tan^2(45° + \phi/2)$

At-rest Earth Pressure Coefficient $K_0 = 1 - \sin \phi$

Extended Marston's theory [23]:

Analytical solutions provided by Aubertin et al. [44], Marston [36] and Terzaghi [40] are for the 2D stope where the fill is subjected to plane strain loading (e.g., a trench). In reality, mine stopes are rarely 2D. Therefore, it is useful to extend these theories to 3D. Therefore, equations were developed for square and circular stopes as well.

Circular stopes are uncommon, but they are quite easy to model using finite element or finite difference methods as axisymmetric problems. Square stopes can only be approximated as axisymmetric problems.

Vertical and horizontal stresses which are acting within the backfilled rectangular stope can be found as follows.

$$\sigma_v = \frac{\gamma w}{2K \tan \delta}\left(\frac{l}{l+w}\right)\left[1 - exp\left\{-2\left(\frac{l+w}{lw}\right)Kz \tan \delta\right\}\right] \tag{1.12}$$

$$\sigma_h = K\sigma_v \tag{1.13}$$

$$\sigma_h = \frac{\gamma w}{2 \tan \delta}\left(\frac{l}{l+w}\right)\left[1 - exp\left\{-2\left(\frac{l+w}{lw}\right)Kz \tan \delta\right\}\right] \tag{1.14}$$

For square stopes ($w = l$),

$$\sigma_v = \frac{\gamma w}{4K \tan \delta}\left[1 - exp\left(-\frac{4Kz \tan \delta}{w}\right)\right] \tag{1.15}$$

$$\sigma_h = \frac{\gamma w}{4 \tan \delta}\left[1 - exp\left(-\frac{4Kz \tan \delta}{w}\right)\right] \tag{1.16}$$

These equations are also valid for circular stopes and storage silos/vessels with circular cross section.

For a very long stope (i.e. $w/l = 0$), Eqs. (1.12) and (1.14) reduce to Eqs. (1.10) and (1.11), given by Aubertin et al. [44].

[Note: Here the backfill material is approximated as dry granular soils (Natural moisture content $w_n = 0$ and cohesion $c = 0$). No pore water pressure is present and hence effective stresses are equal to total stresses].

The reason for using K as K_0 is that typically rock is around two orders of magnitude larger in stiffness than backfill materials, and therefore wall movement is very small and once the backfill is put in place, it would be at rest condition confirming no lateral deformation of the wall due to the fill.

Pirapakaran and Sivakugan [23] have suggested that during loose backfilling, it is suitable to use the wall friction angle (δ) as 2/3 of the backfill friction angle

(ϕ). Numerical solutions have also shown that $K = K_0$ and $\delta = 0.67 \phi$ in the analytical equations give predictions that compare better with the results from the Fast Lagrangian Analysis of Continua (FLAC) simulation program.

Arching theories [36, 40] have suggested that when arching occurs the vertical stress at the bottom of the filled stope is significantly less than that from the self-weight pressure/overburden pressure ($= \gamma z$).

Sivakugan and Widisinghe [32] have developed a general expression for the average vertical normal stress (σ_z) within the mine fill at a depth of 'z' from the top of the fill. For generality, the fill is assumed to have both cohesive and frictional properties, and a uniform surcharge of 'q' is applied at the top of the fill. The fill-wall interface has a friction angle of δ and adhesion of c_a. The lateral earth pressure coefficient is 'K'. The stope cross section has an area 'A' and perimeter 'P'.

$$\sigma_z = \frac{\left(\gamma - \frac{P}{A}c_a\right)}{K(\tan \delta)\frac{P}{A}}\left(1 - e^{-K(\tan \delta)\frac{P}{A}z}\right) + qe^{-K(\tan \delta)\frac{P}{A}z} \tag{1.17}$$

The first component in Eq. (1.17) comes from the fill weight and the second component is from surcharge at the top of the fill.

This Eq. (1.17) can be simplified to Eq. (1.12) by assuming the stope cross section to be rectangular (width $= w$, length $= l$), fill is granular ($c = 0$) and no surcharge is applied ($q = 0$).

The Eq. (1.17) can be simplified to Eq. (1.15) by assuming the stope cross section to be square (width $=$ length $= w$), fill is granular ($c = 0$) and no surcharge is applied ($q = 0$).

The Eq. (1.17) can be simplified to Eq. (1.4) which was originally developed by Marston [36] for plane strain loading situation by assuming a narrow trench where, $l \gg w$, $A = lw$, $P = 2(l + w)$ and $P/A \cong 2/w$ and the backfill is granular ($c = 0$) and no surcharge is applied ($q = 0$).

The Eq. (1.17) can be simplified to Eq. (1.7) which is an extension of the Marston's theory by Terzaghi [40] by assuming a narrow trench situation with plane strain loading, incorporating cohesion to the backfill and no surcharge is applied ($q = 0$).

Concerns with Analytical Equations

Analytical solutions are based on several approximations and simplifications. Two of the common ones are:

(1) The vertical normal stress distribution within the backfill, at any depth, is assumed to be uniform. (i.e., both the analytical and laboratory test solutions reflect the average vertical normal stress at any depth of the fill, although laboratory models treat the soil mass as a particulate medium unlike in analytical approach, and also most numerical models treat the soil mass as a continuum)
(2) Assume an earth pressure coefficient as K_0, K_a or K_p.

1.3.4 Lateral Earth Pressure Coefficient (K) used in Analytical Equations

In a homogeneous fill, the lateral earth pressure coefficient (K) at any given point is defined as the ratio between effective horizontal stress and the effective vertical stress acting on the soil mass.

Marston [36] suggested that the lateral earth pressure coefficient (K) is given by the Rankine's active earth pressure coefficient (K_a) [45].

$$K_a = \tan^2\left(45 - \frac{\phi}{2}\right) = \frac{1 - \sin \phi}{1 + \sin \phi} \tag{1.18}$$

Terzaghi [40] used the lateral earth pressure coefficient given by Krynine [46]. For rough vertical walls, assuming $\delta = \phi$.

$$K = \frac{1}{1 + 2\tan^2 \phi} = \frac{1 - \sin^2 \phi}{1 + \sin^2 \phi} \tag{1.19}$$

Handy [25] suggested that due to rotation of principal stresses near the wall, K increases from K_a towards K_0, and hence suggested the Jaky's [47] expression (Eq. 1.20) for K. Aubertin et al. [44]—modified Marston's theory suggested three cases of lateral earth pressure coefficients as the active (K_a), passive (K_p) (given by Rankine's expressions) and at-rest earth pressure coefficient $(K_0$-from Jaky's expression).

$$K_0 = 1 - \sin \phi \tag{1.20}$$

Relationship between Fill-Wall Interfacial Friction Angle (δ) and Friction Angle of the Fill (ϕ) [32, 48]

The fill-wall interfacial friction angle (δ) can be in the range of $0 - \phi$, depending on the roughness of the wall. In underground mine stopes that are backfilled, the walls formed by blasting and excavation can be very rough (Fig. 1.3). Hence, the slip of the fill takes place along a vertical plane at a few grain sizes away from the rock wall, on a fill-fill interface, where $\delta = \phi$. But in the case of grain silos or storage vessels, walls may not be very rough, and δ may be taken as 0.5–0.7 times ϕ as in the case of pile foundations and retaining walls.

Variation of K tan δ against the Friction Angle of the Fill (ϕ) [32]

For very rough fill-wall interface $(\delta = \phi)$, K tan δ is insensitive to the friction angle of the fill. Here, with the increasing friction angle, tan δ increases, and K decreases such that K tan δ remains approximately the same. Hence, even an estimated friction angle would be adequate when applying analytical equations for a very rough fill-wall interface. Assuming at-rest conditions (i.e., K_0 from Jaky's [47] expression) and

Fig. 1.3 Roughness of mine stope walls

a very rough fill-wall interface (i.e., $\delta = \phi$, underground backfilled mine stopes), $K \tan \delta$ can be approximated as 0.3 for all friction angles ranging from 30° to 49°. For moderately rough fill-wall interface ($\delta = 0.5 \phi$), $K \tan \delta$ increases slightly with the increase in friction angle of the fill.

1.3.5 The Essence of Numerical Modelling

Numerical modelling for mine backfilling is extremely valuable for large underground mines, where field measurements and monitoring of stresses, pore pressures, etc. is usually very challenging, costly or impractical at all. Moreover, software simulations are necessary as a validating tool for laboratory models and also to solve more complex scenarios that cannot be addressed analytically.

Numerical models embedded with correct input parameters, appropriate constitutive models and sensible boundary conditions can give more realistic predictions of stress behaviour within backfilled stopes. Examples are Bloss [49]-CHF TVIS model at Mount Isa mines, Pierce [50]-PF FLAC^{3D} at Brunswick mine, Canada; Rankine et al. [51]-PF FLAC^{3D} at BHP Cannington mine, Australia; Aubertin et al. [44]-HF 2D models using PHASE2, Li et al. [29]-HF 2D models using FLAC which is an explicit finite difference software being used in solving mining and geotechnical problems. It is worthwhile to note that a very long (length is significantly larger) narrow (height is much greater than width) rectangular trench can be modelled as a 2D plane strain problem in FLAC. A circular stope can be modelled representing an axisymmetric problem (symmetric around the center axis). For stopes with square and rectangular cross sections, modelling is in 3D.

Aubertin et al. [44] and Li et al. [29] models:

These models have demonstrated that the vertical normal stress at the bottom of the stope is lower than the actual overburden pressure ($=\gamma z$). Further, the vertical stress reaches a peak value at the mid height of the stope which exceeds the overburden pressure unrealistically. This may be due to the way they placed the fill in the stope during model construction. They assumed that the entire fill is placed instantaneously as one whole body of soil mass which is not the reality. Modelling with layer filling is more sensible and it mimics the actual backfilling process in the mine.

Pirapakaran and Sivakugan [23]:

Through FLAC modelling, it has been investigated that the vertical stress does not exceed the overburden pressure when the stope is filled layer by layer.

FLAC results for the fill stresses from both rectangular and circular stopes compare better with the results from the Marston's analytical equation with $K = K_0$ and $\hat{c} = 0.67\ \phi$. Stress values from Modified Marston's equation [44] with $K = K_0$, $\xi = \phi$ and $K = K_a$, $\delta = \phi$ give lesser equality with the stresses from FLAC results.

Through the analysis of both rectangular and circular stopes, it has been evident that the lateral variation of the vertical normal stress across the stope width is non-uniform and the values are lower near to walls compared to that at the middle region. Also, the reduction of vertical normal stresses along the stope height is significantly increased when compared to the overburden pressures at locations along the stope height. This effect on vertical stresses is known as the arching effect.

It has been shown that the friction angle of the backfill material (ϕ) is increased with the increasing relative density (D_r) of soil mass. In general, the friction angle of any backfill type varies from 30 to 49 degrees, and depends on different packing densities. The influence of friction angles of hydraulic fills on vertical stresses has been studied for narrow stope and circular stope using FLAC simulations. The vertical stresses are approximately the same for all the friction angles within the upper region of the stope and the stresses are nearly equal for friction angles of 30 and 35 degrees in the middle region of the stope, but these are slightly reduced for 40 and 45 degrees. However, vertical stresses differ significantly for all angles closer to the stope floor region and it has showed completely different variations for vertical stresses when compared to analytical solutions. Further research is recommended to understand the stress developments within the stope for various friction angles at stope floor and closer regions.

1.3.6   Stress Variation within Mine Stopes Backfilled with Granular Soils

In-situ stress measurements within backfilled mine stopes have been reported in a few studies from past research [52–55]. Because of limited availability of field data, laboratory and numerical modelling of backfill systems has been carrying out for past few decades [4, 23, 29, 31, 32, 43, 44, 56–62].

Stress developments within backfilled mine stopes should be investigated considering the arching effect. There are three main independent approaches used extensively in geotechnical studies namely analytical equations, laboratory modelling, and numerical/statistical modelling simulations.

Analytical and numerical models based on continuum modelling approach, suggest that vertical normal stress within the fill reaches a maximum value at a certain depth beyond which the stress remains constant, whereas laboratory model tests show that the vertical normal stresses increase continuously even at very large depths without reaching any asymptotic value indicating that the behaviour of the soil mass should be analysed in particulate level.

Given the significance to the particulate modelling in geotechnical engineering, To and Sivakugan [63, 64] has analysed the stress distribution of granular material settling in silos using the Discrete Element Method (DEM). According to To and Sivakugan [64], it concluded that the experimental results for the stresses in the granular fill correlate better with DEM simulation results compared to other analytical and continuum modelling methods as shown in the dimensionless plot in Fig. 1.4. Fill stresses from DEM show a similar trend to that of lab model results where the stresses are increasing with the fill depth without reaching an asymptotic value in the case of analytical and Finite Difference Method (FDM) modelling. DEM results overestimate the vertical stress from the lab model results although the rates are similar. The difference is assumed to be attributed to the use of uniform spherical particles in DEM simulations where the friction is underestimated due to less number of contacts and contact areas, and hence higher stress generation within the fill. Therefore, further research on DEM simulations of stopes/silos filled with granular material considering the angularity of grains with more number of particles is highly recommended.

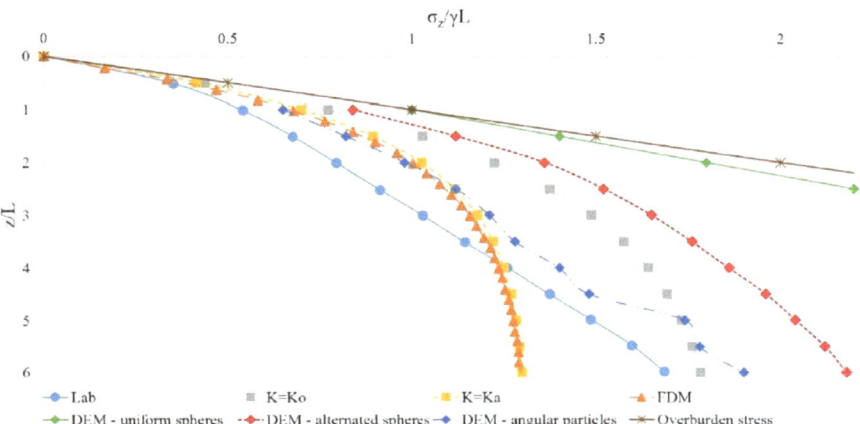

Fig. 1.4 DEM simulation of average vertical stress within a granular fill (*Source* To and Sivakugan [64])

Table 1.1 Evolution of analytical solutions for determining vertical stresses within backfills

Proposer	Material properties	Application	Result	K value	δ
Janssen [41]	Granular	Corn-filled silo	Average vertical normal stress		
Marston [36]	Granular (e.g., hydraulic fills)	2D plane strain loading (e.g., Trenches—very long narrow stopes)	Average vertical normal stress	K_a from Rankine	Between $1/3 \, \phi$ and $2/3 \, \phi$
Terzaghi [40]	Any material (both granular and cohesive)	2D plane strain loading (e.g., Trenches—very long narrow stopes)	Average vertical normal stress	K from Krynine	Equal to ϕ
Aubertin et al. [44]-Modified Marston's theory	Granular (e.g., hydraulic fills)	2D plane strain loading (e.g., Trenches—very long narrow stopes)	Average vertical normal stress	Three cases; K_a and K_p from Rankine's expressions K_0 from Jaky's expression	Equal to ϕ
Prapakaran and Sivakugan [31]	Any material (both granular and cohesive)	3D mine fill stopes and storage vessels with rectangular, square and circular cross sections	Average vertical normal stress	K_0 from Jaky	Equal to $2/3 \, \phi$
Handy [25], Li and Aubertin [59], Singh et al. [65]	Any material (both granular and cohesive)	2D inclined mine stopes	Maximum vertical normal stress	K_0 from Jaky	Equal to $2/3 \, \phi$

The evolution of analytical solutions accounting the arching theory in determining vertical normal stresses within material filled structures is shown in Table 1.1.

1.3.7   Laboratory Modelling of Backfilled Stopes

According to Jayakodi et al. [66], a laboratory model apparatus was used to study the arching effect on the stresses within the hydraulic fill. Figure 1.5a shows a photograph of the laboratory model stope set up called the 'arching apparatus' and it is schematically drawn in Fig. 1.5b. High strength steel wires were used to suspend the model to which a load cell was connected from the top of the frame. The model is positioned vertically such that the minimum possible gap between the stope and balance becomes 0.5 mm, ensuring that there is no any load registered on the balance. When the model is filled in layers with dry hydraulic fill of weight $(m+n)$, a fraction of the fill load is recorded on the balance at the bottom and the rest is registered on the load cell as the fill load carried by the walls. The load cell records the fill load transferred to the wall (m) and the balance reflects the fill load transferred to the bottom (n), given that the balance and load cell readings were zero at the start of the test. At any stage of filing, m and n can be measured separately through this setup. A square shape stope with breath $(B) = 150$ mm and height $= 900$ mm with an open bottom, was tested and the wall condition was varied from rough to smooth, to simulate underground mine wall roughness or a grain silo. Since the stope walls are very rough after blasting, the interfacial friction angle (δ) can be taken as the friction angle of the fill (ϕ) as the failure occurs on the fill-fill interface, not the fill-wall interface. Hence, a similar condition was created using a very rough sandpaper attached to the stope walls of the model, and acrylic planes were used as the smooth walls to represent the condition of a grain silo.

Figure 1.6 shows a dimensionless analysis to compare the laboratory result to that of Marston's equation (Eq. 1.4). Marston's analytical equation was used to express the stress variation for both K_a and K_0 cases separately. The ratio of vertical stress to the product of unit weight and the stope width/breath (i.e., $\sigma/\gamma B$) is plotted against

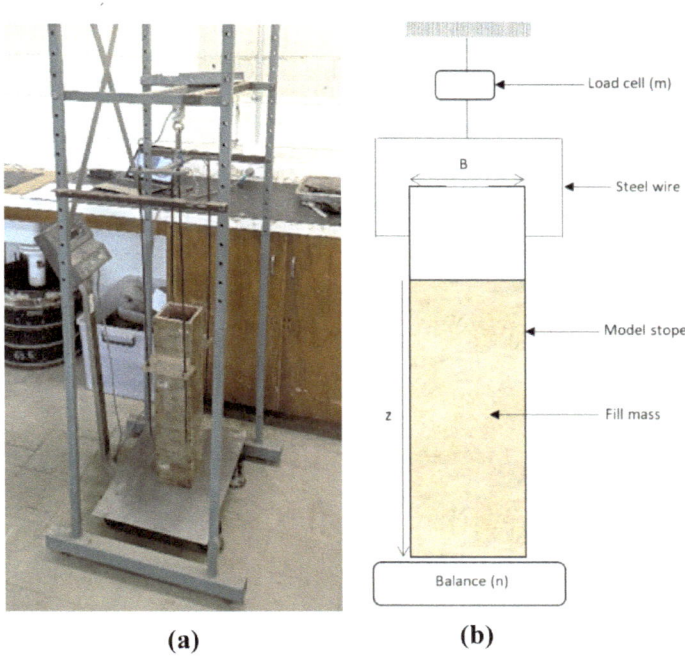

| (a) | (b) |

Fig. 1.5 Laboratory model stope setup **a** photograph **b** schematic diagram

the fill depth and stope breath ratio (i.e., z/B). The dimensionless analysis allows for the comparison of varying stope sizes and fill properties. Figure 1.6 depicts that the Marston's equation with $K = K_a$ largely overestimates the stresses whereas, $K = K_0$, $\delta = \phi$ condition gives more reasonable approximation to the laboratory results. Hence, the at-rest lateral earth pressure coefficient (K_0) is more reliable in Marston's equation (Eq. 1.4) to predict the stress at a depth approaching $6B$, and this was reiterated by Pirapakaran and Sivakugan [23], however $\delta = 0.67\ \phi$ was applied. The mine stope walls are generally more competent and hence the wall movement is not expected in a backfill system. Therefore, the backfill can be considered more stable suggesting that K_0 is more realistic to use in the Marston's equation. Laboratory results from both rough and smooth wall conditions confirmed that the vertical stress is continuously increasing through to $6B$ depth, however the analytical solutions suggested an asymptotic value for the stress at a depth as low as $4B$.

Sivakugan et al. [2] presented two distinct ways of developing the variation of vertical normal stress within the dry granular fill with the depth of the stope through numerical modelling, which give different values of stresses as shown in the dimensionless plot in Fig. 1.7.

Method 1: Vertical stress at the stope bottom for each fill layer, replicating the laboratory model test above (stress variation is measured only at the stope bottom).

Method 2: Vertical stresses at corresponding depths for each layer added, determined when filling the entire stope is completed.

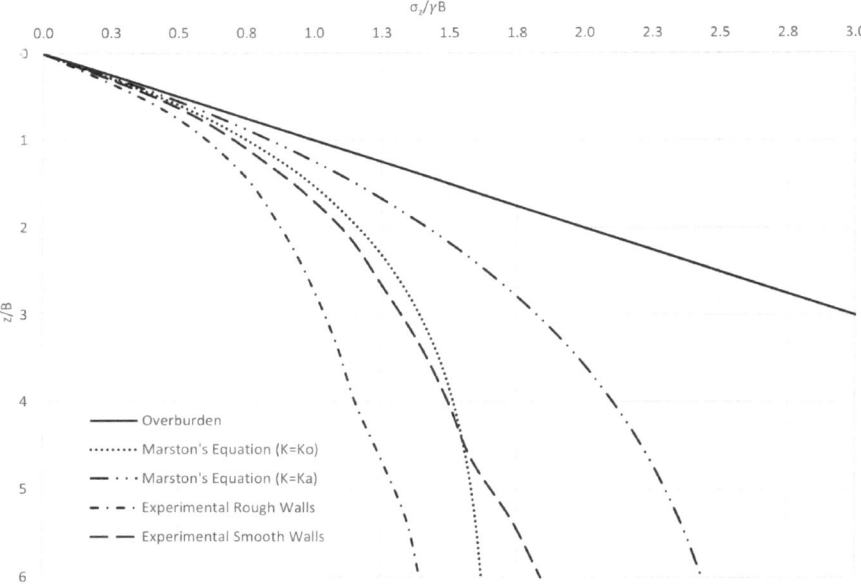

Fig. 1.6 Comparison of laboratory results and Marston's analytical solutions

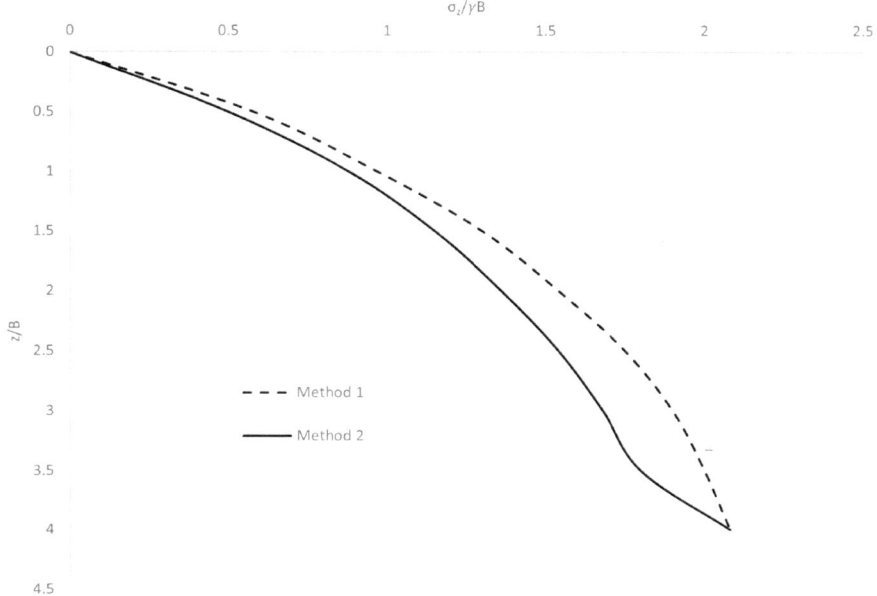

Fig. 1.7 Vertical stresses within granular fill in the stope

According to Sivakugan et al. [2], the two methods give significantly different vertical stress profiles except for the top and the bottom of the stope (see Fig. 1.7). The difference in the stress profiles is attributed to restricting the movement of stope bottom, implying zero displacement in any direction, assumed in numerical modelling.

In Method 2, downward movement of the fill is not restricted for all layers except for the bottom layer and hence, the friction along the wall is fully mobilised in the upper layers. When the fill movement is fully restricted for the bottom layer, more fill load is transferred to the bottom due to partial frictional mobility, which causes a sudden increase in the vertical normal stress closer to the bottom. This behaviour is clearly evident in the numerical modelling work carried out by Fahey [56], Kuganathan [57], Li and Aubertin [4], Pirapakaran and Sivakugan [23], and Ting et al. [67]. This approach is useful to estimate the vertical normal stresses at any depth of the fill, once the filling process has been completed.

In Method 1, the vertical normal stress is determined only at the stope bottom when the fill accumulates, unlike in Method 2. Therefore, the effect of the zero displacement of the stope bottom (i.e., partial mobilisation of shear stresses) is accounted for every layer, and hence, the sharp increase in vertical stress near the bottom, observed in Method 2, is not seen in Method 1. This method is more suitable in situations where the stresses at the bottom of the stope are required throughout the stope filling. For example, from engineering perspective, it is more important to determine the vertical stresses near the underground structure like barricade or at the bottom of the trench/

stope over the course of filling than knowing the vertical stress profile for the entire stope after filling is completed.

The stress patterns obtained from both methods show similar trends for stopes or trenches with different aspect ratios. In the case of Method 1, all the stress profiles obtained for any aspect ratio fall over an approximately similar path and hence, the use of a single stress curve is accepted. Although, a considerable deviation in stresses is observed with regard to the Marston's equation, Method 1 appears to follow the trend of the analytical (Marston) curve, which does not have the sudden increase in stresses near the stope floor known as 'kink', unlike in the curves from Method 2.

1.4 Loadings on Drive Barricades

1.4.1 General

In underground mines, the horizontal access drives that are connected to the mine stope at different sub-levels, are generally barricaded with strong retaining walls before backfilling the stope. The barricades are designed to be free draining such that they allow the water in the fill to seep through while retaining the fill. The construction of a barricade with special porous bricks is shown in Fig. 1.8. Realistic determination of the vertical stresses within the mine stope and correct estimation of loadings onto barricades can be a successful and correct approach to improving the design of barricades. The failure of the barricade can be catastrophic, with in-rush of wet hydraulic fill into the mines, trapping the miners and machinery. Between 1980 and 1997, 11 barricade failures were recorded at Mount Isa Mines in both HFs and CHFs [18]. In 2000, three fatalities were reported due to barricade failure at the Normandy Bronzewing Mine in Western Australia and later that same year, two barricade brick failures were reported as a result of HF contaminant at Osborne Mine in Queensland, Australia [3].

1.4.2 Design of Fill Barricades in Underground Mines

The construction of mine fill barricades is largely associated with the properties of barricade bricks and the stress state of the fill material in the stope and drive. According to Thompson et al. [68], it is anticipated that the horizontal normal stress acting on the barricade below 100 kPa will have a great chance to avoid failure. However, a proper evaluation of the stress state along the drive is essential to carry out safe barricade design work and also to determine the appropriate filling schedule. Analysing barricade stress behaviour for design and construction purposes by means of analytical equations, laboratory tests, and validating with numerical modelling is significantly important for safe and sustainable mining practice.

Fig. 1.8 Barricade being constructed using specially made very porous bricks

Analytical equations are useful for rational assessment of barricade performance for fill loads, however, the solutions are constrained by assumptions for barricade-rock interface properties, geometry simplifications, etc. Unlike in some analytical models that consider the arching of the fill load across the entire barricade, field observations indicate that the likelihood of barricade failure with punching shear at the center area of the wall is high [12, 18]. Kuganathan [69] expresses an analytical equation to determine the stresses on fill barricade as below.

$$\sigma_b = \sigma_o \times exp\left(-\frac{PLK_0 \tan \phi}{A}\right) \tag{1.21}$$

where, σ_b the horizontal stress at the barricade, σ_o the horizontal stress at the drive entrance, P the drive perimeter, A the barricade cross sectional area, L the barricade offset distance, and ϕ the friction angle of the fill.

1.4.3 Laboratory Modelling of Stress State at Drive Barricade

Jayakodi et al. [66] used, a laboratory model barricade comprising 310 mm diameter vertical stope and several drive attachments in the model tests (Fig. 1.9). The granular soil mass in the model has been subjected to surcharge loads (q) ranging from 30 to 900 kPa, and the soil pressures were recorded at the stope centre, stope edge and on the barricade face using three Earth Pressure Cells (EPC) attached in the model. The fill height within the vertical stope was 450 mm. A schematic diagram of the testing apparatus is shown in Fig. 1.10, clearly showing the three EPC locations.

Square (SD) and circular (CD) cross section drives of 100 mm and 150 mm width have been tested using the apparatus shown in Fig. 1.8 to determine the effect of drive shape and size on vertical and horizontal stress distribution. The barricade offset or

Fig. 1.9 a Drive barricade
model apparatus **b** EPCs
attached to the stope base

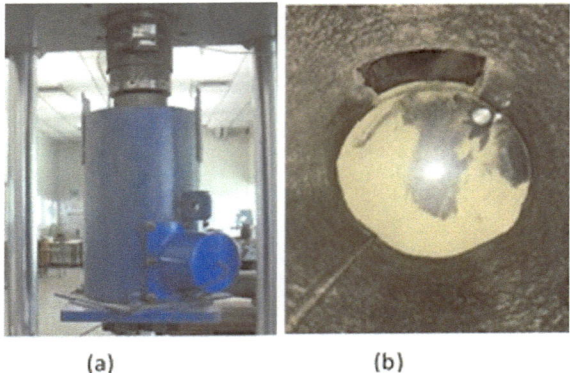

(a)   (b)

Fig. 1.10 Schematic
diagram of the barricade
testing apparatus

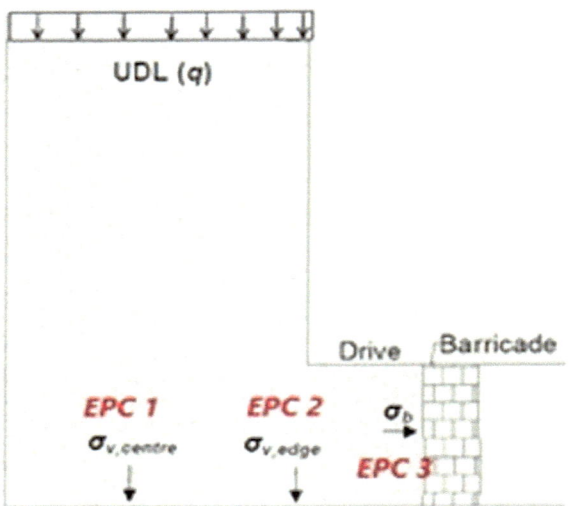

setback distance (L) has been set at four different positions (0, 25, 50, 75 mm) for
each drive size and shape to determine the stress function as offset increases.

Figure 1.11 displays the recorded barricade pressure at each offset tested for the
100 mm drive with circular cross-section (CD100). The plot displays a definite correlation between barricade pressure and offset distance. As the barricade is constructed
at an increasing distance from the stope brow, the stress reduces. This relationship
has been apparent for all four offsets and it confirms that the reduction of lateral load
acting on the barricade with increasing offset distance, encouraging miners to install
the barricade as far as possible from the stope brow. However, the common industry
practice is to keep the barricade offset distance equal to the drive height to reduce
the drainage time of the fill [4].

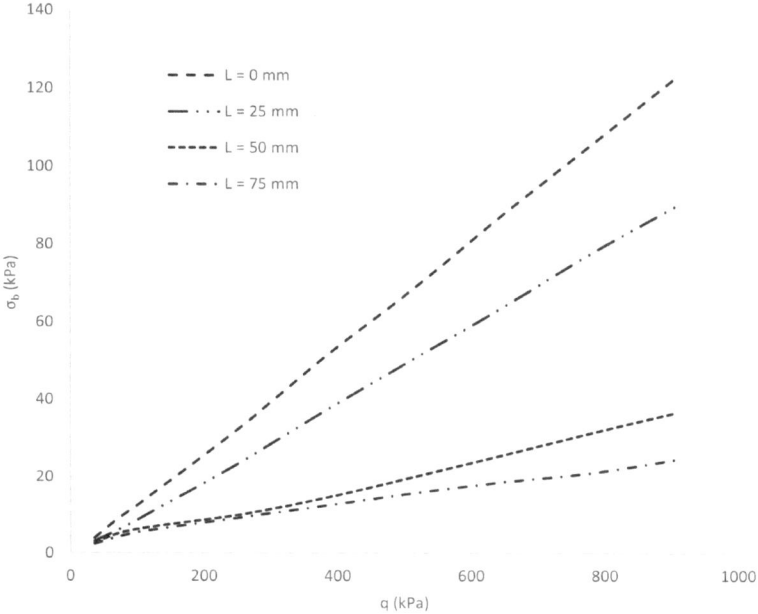

Fig. 1.11 Barricade stress variation with surcharge pressure q for different offset distances L, for CD100

Laboratory results were compared with the Kuganathan's equation (Eq. 1.21) in a dimensionless analysis as shown in Fig. 1.12. The analytical solutions show an overestimation of barricade stresses at an offset (L) approaching the drive height (D) for both active and at-rest earth pressure coefficients, however, the at-rest condition is more comparable with the experimental results justifying the same behaviour as in Fig. 1.6. Furthermore, Kuganathan solutions also verified that the trend of barricade stress reduction with offset distance is non-linear.

A novel laboratory model of square-shaped drive barricade apparatus (Fig. 1.13) has been built recently at James Cook University, Australia to study the arching within drives and to analyse the stress developments on the fill barricade. The vertical stope model apparatus (the 'arching apparatus' referred in Sect. 3.7) will be integrated with the new barricade apparatus to understand the behaviour of fill stresses in a full-scale mine stope through laboratory model simulations. The locations of the soil pressure sensors in the stope are shown in Fig. 1.14.

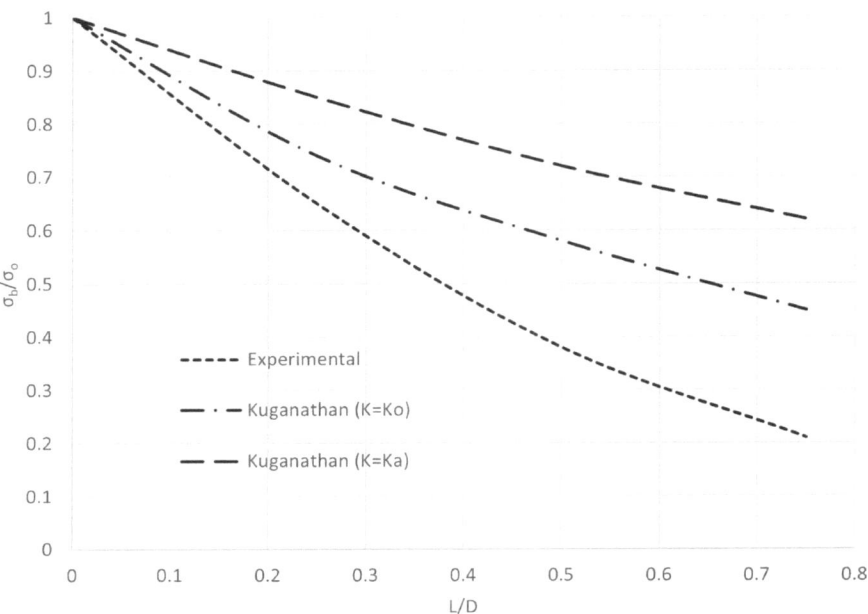

Fig. 1.12 Analytical and experimental comparison of barricade stresses

1.4.4 General Properties of Barricade Bricks

Drive barricades are constructed either using special porous bricks or shotcrete method with drain pipes in place to allow drainage. The porous bricks are composed of gravel, sand, cement and water in the approximate ratio of 40:40:5:1 respectively by weight. It is believed by mining professionals that the porous bricks used for underground barricade construction are susceptible to strength variabilities due to rheological properties of the backfill slurry [18, 70]. The porous bricks with average porosity values of 18–24% and specific gravity values of 2.39–2.59 are designed such that the barricade facilitates free draining of any excess water in the fill. The extensive laboratory testing carried out by Berndt et al. [70] have concluded that the bricks have permeability values significantly higher than that of the hydraulic fill. Typical permeability values of hydraulic fill vary from 2×10^{-4} to 35×10^{-4} cm/s while the permeability for barricade bricks ranges between 1.2×10^{-2} and 3.1×10^{-1} cm/s, which confirms that porous bricks are 2–3 orders of magnitude more permeable than the fill. It has been concluded that the drainage of the backfill system is not influenced by the brick permeability and hence, assuming that the barricade is not affecting the pore pressure development within the fill for numerical modelling exercises is reasonable. Porous bricks that were specially made for the use of barricade construction are shown in Fig. 1.15.

Fig. 1.13 Newly built
square stope-drive barricade
laboratory model apparatus

Fig. 1.14 Inside view of the model stope and locations of soil pressure sensors

Fig. 1.15 Special porous barricade bricks

1.5 Summary

Mine tailing management through mine backfilling is an important part of the underground mining operation which should be done as the mining progresses. Backfilling provides numerous benefits for sustainable mining. Geotechnical aspects of mine backfilling including properties of different backfill types with more focus on hydraulic fills and paste fills, stress developments within backfilled stopes, and the stresses on drive barricades were discussed extensively in this chapter. An accurate estimation of the stress state within the full mine stope is critically important to design safe and competent barricades and to prevent potential barricade failures. Behaviour of a backfill system has been researched over the past few decades through analytical modelling and laboratory modelling. Numerical model simulations have become popular nowadays in an attempt to study the backfill system conveniently. Most of the research on backfill modelling conducted thus far are based on continuum modelling and are limited to vertical stopes. Hence, the particulate approach for backfill modelling is highly recommended and further research on backfill stresses within drive and on barricade are highly encouraged for better designing of barricades. The comprehensive review on geotechnical considerations of mine backfilling and the research based solutions for geotechnical issues and challenges arising when designing a safe backfill system, presented in this chapter will be useful to many stakeholders including researchers, academics, industry professionals and policy makers.

References

1. Minerals Council of Australia. Retrieved May 2, 2023, from https://minerals.org.au/resources/mcas-2020-21-pre-budget-submission/.
2. Sivakugan, N., Widisinghe, S., & Wang, V. Z. (2014). Vertical stress determination within backfilled mine stopes. *International Journal of Geomechanics, 14*(5), 06014011.
3. Grice, T. (2001). Recent minefill developments in Australia. In *7th International Symposium on Mining with Backfill*, Seattle, Washington.
4. Li, L., & Aubertin, M. (2009a). Horizontal pressure on barricades for backfilled stopes. Part I: Fully drained conditions. *Canadian Geotechnical Journal, 46*(1), 37–46.
5. Potvin, Y., Thomas, E., & Fourie, A. (2005). Handbook on mine fill. *Australian Centre for Geomechanics*.
6. Boger, D. V. (1998). Environmental rheology and the mining industry. In *7th International Symposium on Mining with Backfill: Minefill '98*, Australia (pp. 15–17).
7. Rankine, R., Pacheco, M., & Sivakugan, N. (2007). Underground mining with backfills. *Soils and Rocks, 30*(2), 93–101.
8. Piciullo, L., Storrøsten, E. B., Liu, Z., Nadim, F., & Lacasse, S. (2022). A new look at the statistics of tailings dam failures. *Engineering Geology, 303*, 106657.
9. Bloss, M. L. (2014). An operational perspective of mine backfill. In *Minefill 2014: 11th International Symposium on Mining with Backfill* (pp. 15–30).
10. Chileshe, M. N., Syampungani, S., Festin, E. S., Tigabu, M., Daneshvar, A., & Odén, P. C. (2020). Physico-chemical characteristics and heavy metal concentrations of copper mine wastes in Zambia: Implications for pollution risk and restoration. *Journal of Forestry Research, 31*(4), 1283–1293.

11. Nyenda, T., Gwenzi, W., & Jacobs, S. M. (2021). Changes in physicochemical properties on a chronosequence of gold mine tailings. *Geoderma, 395,* 115037.
12. Rankine, K., Sivakugan, N., & Cowling, R. (2006). Emplaced geotechnical characteristics of hydraulic fills in a number of Australian mines. *Geotechnical and Geological Engineering, 24*(1), 1–14.
13. Sivakugan, N., Rankine, R. M., Rankine, K. J., & Rankine, K. S. (2006). Geotechnical considerations in mine backfilling in Australia. *Journal of Cleaner Production, 14*(12–13), 1168–1175.
14. Pettibone, H. C., & Kealy, C. D. (1971). Engineering properties of mine tailings. *Journal of the Soil Mechanics and Foundations Division, ASCE, 97*(SM9), 1207–1225.
15. Herget, G., & De Korompay, V. (1978). In-situ drainage properties of hydraulic backfills. *Proceedings of Mining with Backfill, Research and Innovations, CIM Special, 19,* 117–123.
16. Grice, T. (1998). Underground mining with backfill. In *2nd Annual Summit-Mine Tailings Disposal Systems* (pp. 1–14).
17. Brady, A. C., & Brown, J. A. (2002). Hydraulic fill at Osborne mine. In *Proceedings of the 8th Underground Operators' Conference*, Townsville, Australia (pp. 161–165).
18. Kuganathan, K. (2001). Mine backfilling, backfill drainage and bulkhead construction–A safety first approach. *Australian Mining Monthly* (February), 58–64.
19. Pullum, L. (2003). Pipeline performance. In *International Seminar on Paste and Thickened Tailings, Australian Center for Geomechanics* (pp. 1–13).
20. Robinsky, E. I. (1975). Thickened discharge—A new approach to tailings disposal. *CIM Bulletin, 68,* 47–53.
21. Jewell, R. J., Fourie, A. B., & Lord, E. R. (2002). Paste and thickened tailings-a guide. *Australian Center for Geomechanics,* 152.
22. Clough, G. W., Sitar, N., Bachus, R. C., & Rad, N. S. (1981). Cemented sands under static loading. *Journal of Geotechnical Engineering Division, ASCE, 107*(6), 799–817.
23. Pirapakaran, K., & Sivakugan, N. (2007). Arching within hydraulic fill stopes. *Geotechnical and Geological Engineering, 25*(1), 25–35.
24. Dalvi, R. S., & Pise, P. J. (2008). Effect of arching on passive earth pressure coefficient. In *Proceedings of 12th IACMAG Conference*, Goa, India (pp. 236–243).
25. Handy, R. L. (1985). The arch in soil arching. *Journal of Geotechnical Engineering, 111*(3), 302–318.
26. Take, W. A., & Valsangkar, A. J. (2001). Earth pressures on unyielding retaining walls of narrow backfill width. *Canadian Geotechnical Journal, 38*(6), 1220–1230.
27. Caceras, C. A. (2005). *Effect of delayed backfill on open stope mining methods.* MASc Thesis, University of British Columbia, Vancouver, Canada.
28. Iglesia, G. R., Einstein, H. H., & Whitman, R. V. (1999). Determination of vertical loading on underground structures based on an arching evolution concept. In C. Fernandez & R. A. Bauer (Eds.), *Geo-Engineering for underground facilities* (pp. 495–506). Geo-Institute of ASCE.
29. Li, L., Aubertin, M., Simon, R., Bussière, B., & Belem, T. (2003). Modeling arching effects in narrow backfilled stopes with FLAC. In *3rd International Symposium on FLAC and FLAC3D Numerical Modelling in Geomechanics*, Ontario, Canada (pp. 211–219).
30. Ladanyi, B., & Hoyaux, B. (1969). A study of the trap-door problem in a granular mass. *Canadian Geotechnical Journal, 6*(1), 1–14.
31. Pirapakaran, K., & Sivakugan, N. (2007). A laboratory model to study arching within a hydraulic fill stope. *Geotechnical Testing Journal ASTM, 30*(6), 496–503.
32. Sivakugan, N., & Widisinghe, S. (2013). Stresses within granular materials contained between vertical walls. *Indian Geotechnical Journal, 43*(1), 30–38.
33. Bosscher, P. J., & Gray, D. H. (1985). Soil arching in sandy slopes. *Journal of Geotechnical Engineering, ASCE, 112*(6), 626–635.
34. Low, B. K., Tang, S. K., & Choa, V. (1994). Arching in piled embankments. *Journal of Geotechnical Engineering, ASCE, 120*(11), 1917–1937.
35. Shelke, A., & Patra, N. R. (2008). Effect of arching on uplift capacity of pile groups in sand. *International of Geomechanics, ASCE, 8*(6), 347–354.

36. Marston, A. (1930). The theory of external loads on closed conduits in the light of the latest experiments. *Bulletin 96 of the Iowa Engineering Experiment Station,* 138–170.
37. McCarthy, D. F. (1988). *Essentials of soil mechanics and foundations: Basic geotechnics.* Prentice Hall.
38. Shukla, S. K., & Sivakugan, N. (2013). Load coefficient for ditch conduits covered with geosynthetic-reinforced granular backfill. *International Journal of Geomechanics, 10*(106), 76–82.
39. Spangler, M. G. (1962). Culverts and conduits. In G. A. Leonards (Ed.), *Foundation engineering* (pp. 965–999). McGraw-Hill
40. Terzaghi, K. (1943). *Theoretical soil mechanics* (pp. 66–99). Wiley and Sons.
41. Janssen, H. (1895). Versuche uber Getreidedruck in Silozellen. *Zeitschrift des Vereines deutscher Ingenieure, 39,* 1045–1049.
42. Walters, J. K. (1973). A theoretical analysis of stresses in silos with vertical walls. *Chemical Engineering Science, 28,* 13–21.
43. Mitchell, R., Olsen, R., & Smith, J. (1982). Model studies on cemented tailings used in mine backfill. *Canadian Geotechnical Journal, 19*(1), 14–28.
44. Aubertin, M., Li, L., Arnoldi, S., Belem, T., Bussière, B., Benzaazoua, M., & Simon, R. (2003). Interaction between backfill and rock mass in narrow stopes. In *12th Pan American Conference on Soil Mechanics and Geotechnical Engineering* (pp. 1157–1164).
45. Rankine, W. J. M (1857). On the stability of loose earth. *Philosophical Transactions of the Royal Society, 147,* 9–27. http://doi.org/10.1098/rstl.1857.0003
46. Krynine, D. P. (1945). Discussion of stability and stiffness of cellular cofferdams by K. Terzaghi. *ASCE Transactions, 110,* 1175–1178.
47. Jaky, J. (1944). The coefficient of earth pressure at rest. *Journal of the Society of Hungarian Architects and Engineers, 7*(2), 355–358.
48. Singh, S., Sivakugan, N., & Shukla, S. K. (2010). Can soil arching be insensitive to phi? *International Journal of Geomechanics, ASCE, 10*(3), 124–128.
49. Bloss, M. L. (1992). *Prediction of cemented rock fill stability design procedures and modelling techniques.* Ph.D. Thesis, University of Queensland, Brisbane.
50. Pierce, M. E. (2001). Stability analysis of paste back fill exposes at Brunswick mine. In *Proceedings of 2nd International FLAC Symposium,* Lyon, France (pp. 147–156).
51. Rankine, R. M., Rankine, K. J., Sivakugan, N., Karunasena, W., & Bloss, M. (2001). A numerical analysis of the arching mechanism in paste fills throughout a complete mining sequence. *Proceedings of the 1st Asian Pacific Congress on Computational Mechanics* (pp. 461–466).
52. Belem, T., Harvey, A., Simon, R., & Aubertin, M. (2004). Measurement and prediction of internal stresses in an underground opening during its filling with cemented fill. In *5th International Symposium on Ground Support in Mining and Underground Construction* (pp. 619–630).
53. Knutsson, S. (1981). Stresses in the hydraulic backfill from analytical calculations and in situ measurements. In *Application of Rock mechanics to Cut and Fill mining* (pp. 261–268). Institution of Mining and Metallurgy.
54. Thompson, B. D., Grabinsky, M. W., Bawden, W. F., & Counter, D. B. (2009). In situ measurements of cemented paste backfill in long-hole stopes. In M. Diederichs & G. Grasselli (Eds.), *ROCKENG09; 3rd CANUS Rock Mechanics Symposium,* Toronto.
55. Thompson, B. D., Bawden, W. F., & Grabinsky, M. W. (2012). In situ measurements of cemented paste backfill at the Cayeli Mine. *Canadian Geotechnical Journal, 49*(7), 755–772.
56. Fahey, M., Helinski, M., & Fourie, A. (2009). Some aspects of the mechanics of arching in backfilled stopes. *Canadian Geotechnical Journal, 46*(11), 1322–1336.
57. Kuganathan, K. (2005). Hydraulic fills. In Y. Potvin, E. D. Thomas, & A. Fourie (Eds.), *Handbook on Mine Fill* (pp. 23–47).
58. Li, L., Aubertin, M., & Belem, T. (2005). Formulation of a three dimensional analytical solution to evaluate stresses in backfilled vertical narrow openings. *Canadian Geotechnical Journal, 42*(6), 1705–1717.

59. Li, L., & Aubertin, M. (2008). An improved analytical solution to estimate the stress state in subvertical backfilled stopes. *Canadian Geotechnical Journal, 45*(10), 1487–1496.
60. Mitchell, R., & Roettger, J. (1984). Bulkhead pressure measurements in model fill pours. *CIM Bulletin, 77*(888), 50–55.
61. Ting, C. H., Sivakugan, N., & Shukla, S. K. (2012). Laboratory simulation of the stresses within inclined stopes. *Journal of Geotechnical Testing, 35*(2), 1–15.
62. Widisinghe, S. D. (2014). *Stress developments within a backfilled mine stope and the lateral loading on the barricade.* Ph.D. thesis, James Cook University.
63. To, P., & Sivakugan, N. (2018a). Arching of granular flow under loading in silos. In *Proceedings of the 9th European Conference on Numerical Methods in Geotechnical Engineering (NUMGE)* (pp. 857–862). CRC Press.
64. To, P., & Sivakugan, N. (2018b). Boundary stress distribution in silos filled with granular material. In *Proceedings of the 9th European Conference on Numerical Methods in Geotechnical Engineering (NUMGE)* (pp. 863–868). CRC Press.
65. Singh, S., Shukla, S., & Sivakugan, N. (2011). Arching in inclined and vertical mine stopes. *Geotechnical and Geological Engineering, 29*(5), 685–693.
66. Jayakodi, J. D. S. U., Bennett, R. J., Reddicliffe, A. J., Sivakugan, N., & To, P. (2021). Laboratory modelling of stresses within the minefills and loads on barricades. In *Proceedings of 3rd International Conference in Geotechnical Engineering*, Colombo, Sri Lanka (pp. 247–252).
67. Ting, C. H., Shukla, S. K., & Sivakugan, N. (2011). Arching in soils applied to inclined mine stopes. *International Journal of Geomechanics, 11*(1), 29–35.
68. Thompson, B. D., Hunt, T., Malek, F., Grabinsky, M. W., & Bawden, W. F. (2014). In situ behaviour of cemented hydraulic and paste backfills and the use of instrumentation in optimising efficiency.
69. Kuganathan, K. (2002). A method to design efficient mine backfill drainage systems to improve safety and stability of backfill bulkheads and fills. In *8th Underground Operators' Conference, Growing Our Underground Operations* (pp. 181–188).
70. Berndt, C. C., Rankine, K. J., & Sivakugan, N. (2007). Materials properties of barricade bricks for mining applications. *Geotechnical and Geological Engineering, 25*(4), 449–471.
71. Geoscience Australia. Retrieved Sept 15, 2021, from https://www.ga.gov.au/education/classroom-resources/minerals-energy/australian-mineral-facts.

Chapter 2
CPTu-Based Soil Behaviour Type Indexes that are Independent of Sleeve Friction Readings: An Application in Tailings

Luis Alberto Torres-Cruzⓘ**, Nico Vermeulen, and Abideen Owolabi**ⓘ

2.1 Background

The piezocone penetration test (CPTu) is widely used for the stratigraphic and mechanical characterization of soil deposits including human-made fills such as tailings (e.g., [1, 2]). The CPTu involves pushing an instrumented probe into a soil deposit at a constant rate, while taking readings of cone tip resistance (q_c), sleeve friction (f_s), and dynamic pore water pressure (u_2) [3, 4]. These three readings are typically supplemented by the equilibrium pore water pressure (u_0) measured during dissipation tests [1].

One important application of the CPTu is to assess soil behaviour type (SBT). Generally, this process relies on two-dimensional SBT charts that have normalised CPTu readings on their axes. SBT charts are divided into regions that correspond to different soil behaviour types (e.g., sand, clay, etc.) or different states (e.g., dilative vs contractive). Herein we focus on the classification into different soil behaviour types. It is worth pointing out that, in this regard, the outcome of an SBT chart is not generally quantitative but rather qualitative. That is, the outcome is not a number but a category such as sand, silt, clay, etc.

The implementation and interpretation of SBT charts is greatly facilitated by the use of SBT indexes capable of quantifying the classification performed by the chart [5]. In this context, an SBT index is a function of the normalised cone parameters that plot along the two axes of the SBT chart. The key characteristic of an SBT index function is that its contours of constant value coincide, at least approximately, with the boundaries of the different soil type regions of the chart. The main benefit of a

L. A. Torres-Cruz (✉) · A. Owolabi
University of the Witwatersrand, Johannesburg, South Africa
e-mail: LuisAlberto.TorresCruz@wits.ac.za

N. Vermeulen
Jones and Wagener (PTY) LTD, Johannesburg, South Africa

S. K. Das et al. (eds.), *Geoenvironmental and Geotechnical Issues of Coal Mine Overburden and Mine Tailings*, Springer Transactions in Civil and Environmental Engineering, https://doi.org/10.1007/978-981-99-6294-5_2

quantitative SBT index, as opposed to a qualitative soil type descriptor, is that an SBT index can conveniently be used in calculations to derive soil parameters (e.g., [5, 6]).

Perhaps the best known SBT index is I_c of which there are two widely used versions. One proposed by Jefferies and Davies [5] and another one proposed by Robertson and Wride [6]. Herein we use $I_{c\text{-}JD}$ to refer to the former version and $I_{c\text{-}RW}$ to refer to the latter. Equations 2.1 and 2.2 present the functions for both versions of I_c. Hereafter, the equations related to $I_{c\text{-}JD}$ are taken from Jefferies and Been [7] which presents equations that are slightly different from those in Jefferies and Davies [5] (different definition of Q'). And the equations related to $I_{c\text{-}RW}$ are taken from the update presented in Robertson [8].

$$I_{c-RW} = f(Q_{tn}, F) = \sqrt{\left(3.47 - \log_{10} Q_{tn}\right)^2 + \left(1.22 + \log_{10} F\right)^2} \qquad (2.1)$$

$$I_{c-JD} = f(Q', F) = \sqrt{\left(3 - \log_{10} Q'\right)^2 + \left(1.5 + 1.3\log_{10} F\right)^2} \qquad (2.2)$$

Implementation of $I_{c\text{-}RW}$ requires the definition of the following parameters.

$$Q_{tn} = \left(\frac{q_t - \sigma_v}{p_a}\right)\left(\frac{p_a}{\sigma'_v}\right)^n \qquad (2.3)$$

$$n = 0.381 I_{c-RW} + 0.05 \frac{\sigma'_v}{p_a} - 0.15 \qquad (2.4)$$

$$F = \left(\frac{f_s}{q_t - \sigma_v}\right) 100\% \qquad (2.5)$$

And implementation of $I_{c\text{-}JD}$ requires the definition of the following additional parameters.

$$Q' = Q(1 - B_q) + 1 \qquad (2.6)$$

$$Q = \frac{q_t - \sigma_v}{\sigma'_v} \qquad (2.7)$$

$$q_t = q_c + u_2(1 - a) \qquad (2.8)$$

$$B_q = \frac{u_2 - u_0}{q_t - \sigma_v} \qquad (2.9)$$

In the preceding equations σ_v and σ_v' are the total and effective vertical stresses prior to the CPTu, p_a is the atmospheric pressure in units that ensure a dimensionless Q_{tn}, n is a dimensionless stress exponent capped at 1, and a is the unequal area

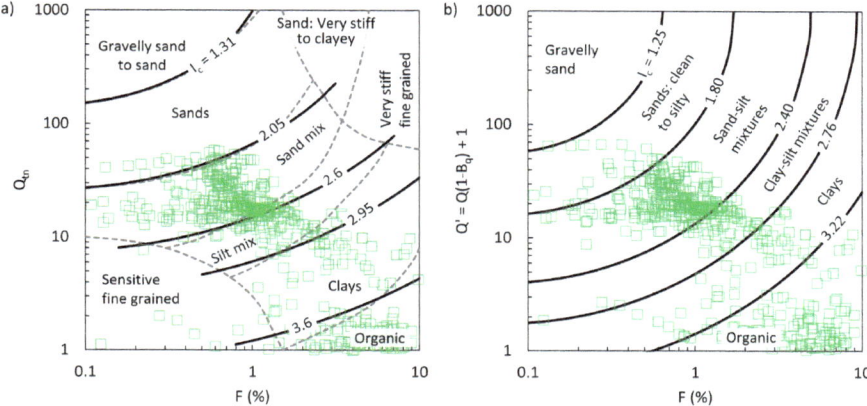

Fig. 2.1 **a** The Q_{tn}-F chart [8] with contours of $I_{c\text{-}RW}$ and **b** the Q'-F chart defined by contours of $I_{c\text{-}JD}$ [7]. Notes: (1) In Fig. 2.1a, the original chart boundaries are shown with dashed lines and $I_{c\text{-}RW}$ contours are shown with continuous lines. (2) Square markers correspond to sounding *CPTU-05-05N_Updated* reported in Robertson et al. [9]

cone factor. The interdependence between $I_{c\text{-}RW}$, Q_{tn}, and n results in an iterative calculation procedure being required to compute these parameters.

Equation 2.1 emphasises that $I_{c\text{-}RW}$ is a function of Q_{tn} and F. This is a direct consequence of Q_{tn} and F being the two variables plotted along the axes of the SBT chart on which $I_{c\text{-}RW}$ is based [6, 8]. Similarly, Eq. 2.2 emphasises that $I_{c\text{-}JD}$ is a function of Q' and F. As such, the contours of $I_{c\text{-}JD}$ can be used to define an SBT chart in Q'-F space [5, 7]. Figure 2.1 shows the SBT charts defined by the contours of both versions of I_c. In the case of $I_{c\text{-}RW}$, the figure also shows the underlying SBT chart whose boundaries are approximated by the contours of $I_{c\text{-}RW}$ (Fig. 2.1a).

Both definitions of I_c were inspired by the Q-F SBT chart proposed by Robertson [10]. As such, both definitions include the normalised sleeve friction F. SBT charts that are independent of sleeve friction f_s (e.g., [11–13]) are also widely used by geotechnical practitioners. But it appears to not be as widely known that the implementation of some of these sleeve friction-independent charts can also be facilitated by SBT indexes. These indexes that do not rely on sleeve friction f_s readings are convenient complements of I_c since investigations suggest that the repeatability of f_s remains a challenge in soft soils [14]. The purpose of this chapter is to illustrate the use of two sleeve friction-independent SBT indexes in the characterisation of tailings. Notwithstanding, the approach presented herein is applicable to all soil deposits in general.

2.2 Sleeve Friction-Independent SBT Charts and Their Corresponding Indexes

Figure 2.2 shows two SBT charts that do not require sleeve friction measurements: one in Q-Bq space [11] and the other in Q-$U2$ space [12, 13]. $U2$ is a normalised measure of excess pore water pressure defined as

$$U2 = \frac{u_2 - u_0}{\sigma'_v} \tag{2.10}$$

Both SBT charts shown in Fig. 2.2 have been used to develop SBT indexes whose contours approximate some of the boundaries of the charts. In the case of the Q-B_q chart (Fig. 2.2a), the corresponding SBT index is $I_{Q\text{-}Bq}$ [16] which is defined with a piecewise function as

$$I_{Q-Bq} = f(Q, B_q) = \begin{cases} Q \cdot 10^{-1.9B_q}, & B_q \geq 0 \\ Q \cdot 10^{2.8B_q}, & B_q < 0 \end{cases} \tag{2.11}$$

For the Q-$U2$ chart (Fig. 2.2b), contours of B_q approximate the boundaries of the undrained (i.e., clay) regions [17]. Furthermore, using results from non-plastic industrial waste, Torres-Cruz and Vermeulen [18] argued that the drained region of the Q-$U2$ chart can also be defined by B_q contours. These two observations imply that B_q is a suitable SBT index for the Q-$U2$ chart [15]. It should be noted that the $B_q = 1$ contour line in the Q-$U2$ chart is not a boundary but rather is intended as a link to

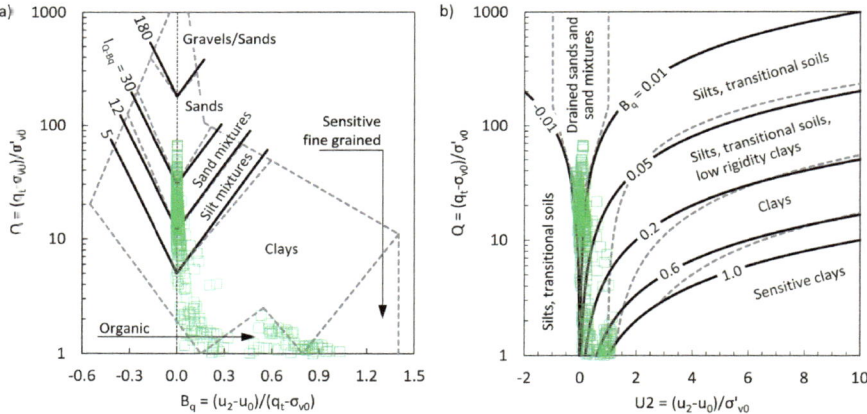

Fig. 2.2 SBT charts that are independent of sleeve friction: **a** the Q-B_q chart [11] with contours of $I_{Q\text{-}Bq}$ and **b** the Q-$U2$ chart [12, 13] with contours of B_q. Notes: (1) The original chart boundaries are shown with dashed lines and SBT index contours are shown with continuous lines. (2) Figure 2.2b adapted from [15]. (3) Square markers correspond to sounding *CPTU-05-05N_Updated* reported in Robertson et al. [8]

Table 2.1 Approximate correspondence between Bq and drainage conditions based on the Q-$U2$ chart proposed by Schneider et al. [12, 13]. *Source* Fourie et al. [15].

Bq range	Drainage conditions
<-0.01	Partial drainage
-0.01 to 0.01	Fully drained
0.01 to 0.05	Partial drainage
0.05 to 0.2	Partial drainage or undrained
>0.2	Undrained

the Q-B_q chart [13]. It is convenient to rewrite B_q in terms of Q and $U2$ to emphasise that B_q is a function of the two parameters plotted on the axes of its underlying SBT chart.

$$B_q = f(Q, U2) = \frac{U2}{Q} \tag{2.12}$$

Accepting that during a CPTu sounding sands remain drained, clays remain undrained, and silts and transitional soils exhibit partial drainage, then the approximate correspondence between drainage conditions and B_q is as given in Table 2.1. It is worth noting that SBT charts or indexes that involve excess pore water pressure ($u_e = u_2 - u_0$) are only applicable in saturated soils and where readings are not affected by cavitation.

2.3 Implementation at a Tailings Deposit

A CPTu sounding from Dam I at the Córrego do Feijão iron mine near Brumadinho, Brazil, is used herein to illustrate the use of the sleeve friction-independent SBT indexes. Dam I exhibited alternating layers of fine and coarse tailings [9]. Figure 2.3 shows profiles of $I_{Q\text{-}Bq}$ and B_q together with profiles of the more common indexes $I_{c\text{-}RW}$ and $I_{c\text{-}JD}$. The profiles of the four indexes reflect the layered structure of the deposit and generally agree on the interpreted stratigraphy. These results highlight the possibility of using $I_{Q\text{-}Bq}$ and B_q to complement $I_{c\text{-}RW}$ and $I_{c\text{-}JD}$.

One practicality of the B_q profile is that, as shown in Fig. 2.3b, the spread of values may be best represented on a logarithmic scale. This precludes the possibility of directly plotting negative B_q values, which instead require a different line style for their identification. Interpretation of the B_q profile must consider that all B_q values smaller (more negative) than 0.01 classify as silts or transitional soils. That is, the Q-$U2$ SBT chart is not symmetric about the $U2 = 0$ axis [12, 13].

Although the profiles in Fig. 2.3 are useful in characterising the stratigraphy of the deposits, there is also value in plotting the data on SBT charts as this provides additional information regarding parameters such as the overconsolidation ratio and sensitivity [10, 13]. To emphasise this point the CPTu data from Dam I is also plotted in the SBT charts shown in Figs. 2.1 and 2.2.

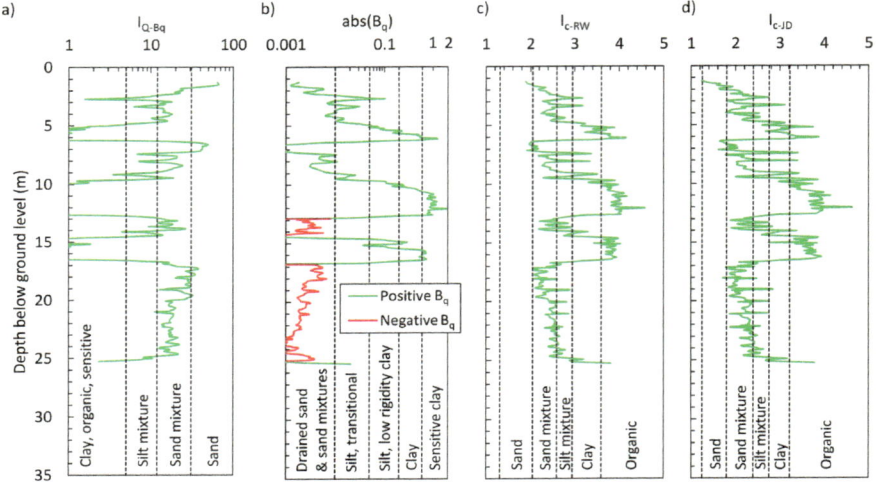

Fig. 2.3 Stratigraphy of Dam I as inferred from profiles of **a** $I_{Q\text{-}Bq}$, **b** B_q, **c** $I_{c\text{-}RW}$, and **d** $I_{c\text{-}JD}$. *Note* Data from sounding *CPTU-05-05N_Updated* reported in Robertson et al. [9]

2.4   Conclusion

The sleeve friction-independent SBT indexes $I_{Q\text{-}Bq}$ and B_q can facilitate the implementation and interpretation of the $Q\text{-}B_q$ and $Q\text{-}U2$ soil behaviour type charts, respectively. The indexes effectively quantify the soil type classification of their corresponding SBT chart which makes the results amenable to mathematical manipulation.

When applied to a CPTu sounding conducted in iron tailings both indexes produced SBT classifications that are comparable to those yielded by the more common indexes $I_{c\text{-}RW}$ and $I_{c\text{-}JD}$. The indexes $I_{Q\text{-}Bq}$ and B_q are independent of sleeve friction readings and are thus particularly useful when there are concerns about the repeatability of such readings. Furthermore, these indexes provide another layer of information to assist with the interpretation of geotechnical characteristics.

References

1. Lunne, T., Powell, J. J. M., & Robertson, P. K. (1997). *Cone penetration testing in geotechnical practice*. Blackie Academic & Professional.
2. Mayne, P. W. (2007). *Cone penetration testing: A synthesis of highway practice. Project 20-5.* Transportation Research Board, Washington, DC NCHRP Synthesis, 368.
3. ASTM D5778-20 (2020). *Standard test method for performing electronic friction cone and piezocone penetration testing of soils*. ASTM International.
4. ISO 22476-1:2022. (2022). *Geotechnical investigation and testing-Field testing-Part 1: Electrical cone and piezocone penetration test.*

5. Jefferies, M. G., & Davies, M. P. (1993). Use of CPTU to estimate equivalent SPT N60. *Geotechnical Testing Journal, 16*(4), 458–468.
6. Robertson, P. K., & Wride, C. E. (1998). Evaluating cyclic liquefaction potential using the cone penetration test. *Canadian Geotechnical Journal, 35*(3), 442–459.
7. Jefferies, M., & Been, K. (2015). *Soil liquefaction: A critical state approach* (Second edition). Taylor & Francis.
8. Robertson, P. K. (2009). Interpretation of cone penetration tests—A unified approach. *Canadian Geotechnical Journal, 46*(11), 1337–1355.
9. Robertson, P. K., de Melo, L., Williams, D. J., & Wilson G. W. (2019). *Report of the expert panel on the technical causes of the failure of Feijao Dam I. Commissioned by Vale.*
10. Robertson, P. K. (1990). Soil classification using the cone penetration test. *Canadian Geotechnical Journal, 27*(1), 151–158.
11. Robertson, P. K. (1991). Soil classification using the cone penetration test: Reply. *Canadian Geotechnical Journal, 28*(1), 176–178.
12. Schneider, J. A., Hotstream, J. N., Mayne, P. W., & Randolph, M. F. (2012). Comparing CPTU Q-F and Q–Δu 2/σv0′ soil classification charts. *Géotechnique Letters, 2*(4), 209–215.
13. Schneider, J. A., Randolph, M. F., Mayne, P. W., & Ramsey, N. R. (2008). Analysis of factors influencing soil classification using normalized piezocone tip resistance and pore pressure parameters. *Journal of Geotechnical and Geoenvironmental Engineering, 134*(11), 1569–1586.
14. Gundersen, A. S., Lindgård, A., & Lunne, T. (2020). *Impact of cone penetrometer type on measured CPTU parameters at 4 NGTS sites. Silt, soft clay, sand and quick clay.*
15. Fourie, A., Verdugo, R., Bjelkevik, A., Torres-Cruz, L. A., & Znidarcic, D. (2022). Geotechnics of mine tailings: a 2022 state of the art. In *Proceedings of the 20th International Conference on Soil Mechanics and Geotechnical Engineering.*
16. Torres-Cruz, L. A. (2015). CPT-based soil type classification in a platinum tailings storage facility. from fundamentals to applications in geotechnics. In *Proceedings of the 15th Pan-American Conference on Soil Mechanics and Geotechnical Engineering*, 15–18 November 2015.
17. Robertson, P. K. (2016). Cone penetration test (CPT)-based soil behaviour type (SBT) classification system—An update. *Canadian Geotechnical Journal, 53*(12), 1910–1927.
18. Torres-Cruz, L. A., & Vermeulen, N. (2018). CPTu-based soil behaviour type of low plasticity silts. In M. A. Hicks, F. Pisano, & J. Peuchen (Eds.), *Cone penetration testing 2018.*

Chapter 3
Assessment of Mine Overburden Dump Stability Using Numerical Modelling

Tarun Kumar Rajak⑩ **and Laxmikant Yadu**⑩

3.1 Introduction

The increased demand for raw materials in thermal power plants led in mineral extraction and a significant expansion in the volume and size of OBD dump. The improper handling of OBD materials may result in dump failure and various difficulties such as decreased mining efficiency, endangering equipment, and workers [1], and adverse environmental effects [2, 3]. Several aspects, including material physico-mechanical properties, geological variables, geometrical factors, and hydraulic factors [4], influence dump stability. External dumping necessitates increasing the height and slope of the dump to achieve maximum dump capacity. However, increasing the geometry of the external dump necessitates a thorough study of the geo-mechanical characteristics of OBD materials [5]. According to the Directorate General of Mines Safety (DGMS), the bench's height shall not exceed 30.0 m. However, given on the geo-mechanical qualities of the available OBD material, an attempt to enhance the individual bench height may be beneficial for maximum dump capacity within the available land area. With increased coal consumption, a large volume of fly ash is also produced, necessitating an alternate manner of safe utilisation. The use of fly ash in mining OBD material may provide an effective form of usage. Several research showed that fly ash might be utilised to alter the engineering qualities of soft soils [6–11]. Fly ash is a desirable construction material for embankments due to its low specific value, high frictional properties, and ease of compaction [12–14]. According to several research, fly ash-OBD mix material may also be employed as

T. K. Rajak (✉)
Department of Civil Engineering, Shri Shankaracharya Institute of Professional Management and Technology, Raipur 492015, India
e-mail: t.rajak@ssipmt.com

L. Yadu
Department of Civil Engineering, National Institute of Technology, Raipur, India

a paving material in mine haul roads [15–20]. According to Cockrell and Leonard [21], Fly ash, when combined with moisture and calcium oxide, produces a cementitious compound that can be used to strengthen the strength of soft soil. Mallick and Mishra [19] examined the strength behaviour of clinker stabilised OBD-fly ash blends and proposed that the mix might be used as a sub-base material in mine haul roads at optimal proportions. Similarly, the applicability of additional materials such as lime [16, 22], and cement [18] for improving the strength qualities of fly ash-OBD material has been investigated.

GGBS was mixed with the fly ash-OBD material in this investigation. When crushed to a finer particle size, GGBS, derived from the iron industry, exhibits cementitious behaviour [23, 24]. It aids in reducing swelling behaviour, flexibility, and improving the physico-mechanical properties of expansive soil [25–27]. According to Sharma and Sivapullaiah [28], combining fly ash with GGBS can be more beneficial than using it alone since it aids in the formation of the calcium-silicate-hydrate gel matrix. However, for optimal usage in the mining area, the strength properties of GGBS stabilised fly ash-OBD material must be studied. It was observed that an optimum GGBS content for the stabilization of the soft soil was found as 6–15% by weight [23–27], thus in the present study, 6, 9, and 12% GGBS was added.

The intent of the study was to look at the viability of using fly ash in conjunction with the OBD material for external mining dumps. The study assessed the compaction and shear strength capabilities of mixes containing varied concentrations of OBD material, fly ash, and GGBS. FLAC/Slope software was used to calculate the FOS of an external dump comprising varied mix proportions and different geometrical configurations. An overall dump height of 120.0 m was investigated using four benches of 30.0 m (height) each, three benches of 40.0 m (height) each, and two benches of 60.0 m (height) each. Further, the slope angle was varied from 28° at 2° interval [29] to suggest the optimum slope angle for each set. A Multiple Linear Regression (MLR) model and Artificial Neural Network (ANN) model has been developed to evaluate the FOS based on various influencing parameters.

3.2 Study Area

The current study focuses on the stability of the external overburden dump in the Mand Raigarh Coalfield in Chhattisgarh, India. The research area and OBD material collecting site are depicted in Figs. 3.1 and 3.2, respectively. Mand Raigarh coalfield is located in the Gondwana basin, which stretches northwest-southeastly (NW–SE). Table 3.1 shows the stratigraphy of the studied region.

Fig. 3.1 Location of Mand
Raigarh coalfield (Study
area)

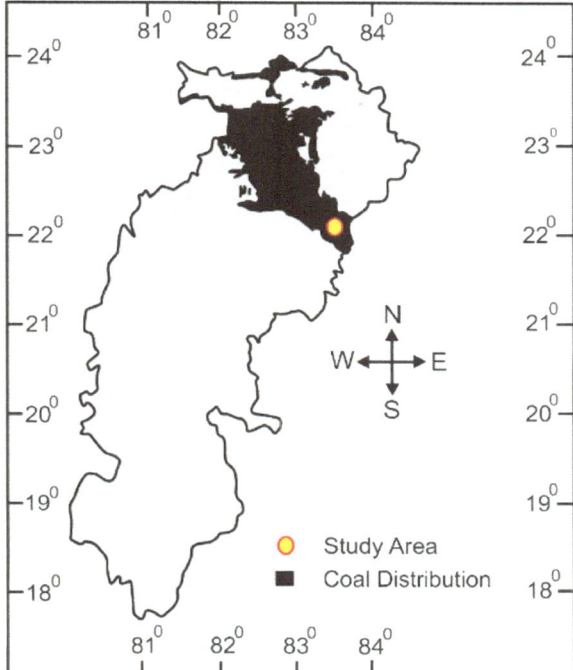

Fig. 3.2 OBD material from
dump site of Mand Raigarh
coalfield

3.3 Materials and Methods

External dump OBD material was obtained from the Gare Pelma Coal mines, Raigarh, Chhattisgarh, India. Fly ash and GGBS were collected from the NSPCL thermal power plant in Bhilai, Chhattisgarh, India, and the Bhilai Steel Plant. The specific

Table 3.1 Lithology and stratification of Mand Raigarh coalfield

Formation	Thickness (m)	Lithology
Recent	1.00–17.00	Alluvium/soil basalt flows and dolerite dykes
Barren measure	3.00–237.97	Fine to medium-grained sandstones shale and intercalation of shale and sandstone, carbonaceous shale and coal seam
Barakar	160.50–482.29	Coarse grain sandstone, medium grain sandstones, shales, and coal seams
Talchir	2.00–8.80	Diamictites, sandstones, shales, rhythmites and turbidites
Basement	2.10–8.25	Metamorphic

gravity of the OBD material, fly ash, and GGBS was determined using ASTM D854-14 [30]. ASTM-D6913M was used to analyse the particle size distribution of the OBD material and fly ash [31]. Energy Dispersive X-ray (EDX) and Scanning Electron Microscope (SEM) tests were used to characterise the chemical and mineral structure of the OBD material, fly ash, and GGBS. To establish the Maximum Dry Density (MDD) and Optimum Moisture Content (OMC) of the OBD material and fly ash, standard and modified Proctor compaction tests were done in accordance with ASTM D698-12, [32] and ASTM D1557 [33], respectively. Samples were created and tested by blending the OBD material, fly ash, and GGBS in various proportions (Table 3.2).

Table 3.2 Sample preparation at different mix proportions

Sample name	Mix proportions containing
90OBD10F	90% OBD and 10% fly ash
84OBD10F6GGBS	84% OBD, 10% fly ash, and 6% GGBS
81OBD10F9GGBS	81% OBD, 10% fly ash, and 9% GGBS
78OBD10F12GGBS	78% OBD, 10% fly ash, and 12% GGBS
80OBD20F	80% OBD and 20% fly ash
74OBD20F6GGBS	74% OBD, 20% fly ash, and 6% GGBS
71OBD20F9GGBS	71% OBD, 20% fly ash, and 9% GGBS
68OBD20F12GGBS	68% OBD, 20% fly ash, and 12% GGBS
70OBD30F	70% OBD and 30% fly ash
64OBD30F6GGBS	64% OBD, 30% fly ash, and 6% GGBS
61OBD30F9GGBS	61% OBD, 30% fly ash, and 9% GGBS
58OBD30F12GGBS	58% OBD, 30% fly ash, and 12% GGBS
60OBD40F	60% OBD and 40% fly ash
54OBD40F6GGBS	54% OBD, 40% fly ash, and 6% GGBS
51OBD40F9GGBS	51% OBD, 40% fly ash, and 9% GGBS
48OBD40F12GGBS	48% OBD, 40% fly ash, and 12% GGBS

3.3.1 Compaction Test

Modified Proctor compaction tests were performed for all mix amounts (Table 3.2) in accordance with ASTM D1557 [33]. To determine the MDD and OMC of all samples, the compaction curve was drawn. The MDD and OMC acquired were utilised to prepare the sample for the triaxial test.

3.3.2 Triaxial Test

All samples were subjected to a triaxial test (unconsolidated undrained test) in accordance with ASTM D2850 [34]. Samples of 38 mm diameter and 76 mm length were made and tested at three confining pressures, namely 49.03, 98.06, and 147.10 kPa, to determine the angle of internal friction and cohesiveness of each material.

3.4 Dump Stability Analysis

The external dump's stability was assessed using numerical modelling with FLAC/Slope software, a finite difference method-based computational tool. FLAC/Slope was used to calculate the dump's factor of safety (FOS), failure pattern, vector velocity, and shear strain rate, offering a quick and efficient method for these computations. FOS is the ratio between actual shear strength to the reduced shear strength presented in Eqs. (3.1) and (3.2) which is determined by Strength reduction techniques. In this method, trial FOS (F') is considered to reduce the material's strength i.e., angle of internal friction (ϕ) and cohesion (c) till failure initiates. F' represents the real FOS in the event of failure.

$$c' = c/F' \tag{3.1}$$

$$\phi' = \arctan(1/F' \tan \phi) \tag{3.2}$$

3.4.1 Simulation of Dump Slope

In this study, overburden dump was simulated in different stages. Model stage and build stage are used to define the geometrical and geomaterial properties of the dump. FOS of the dump was determined in the solve stage by bracketing approach. Further, the failure pattern, velocity vector, plasticity indicator, and shear strain rate are determined in the plot stage.

Fig. 3.3 Numerically simulated OBD dump having four benches of individual height and bench slope 30.0 m and 44° respectively

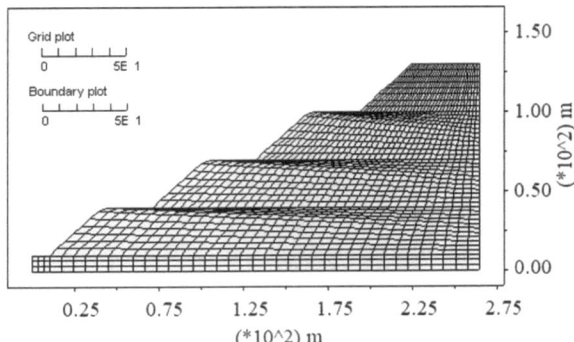

Fig. 3.4 Numerically simulated OBD dump having three benches of individual height and bench slope of 40.0 m and 38° respectively

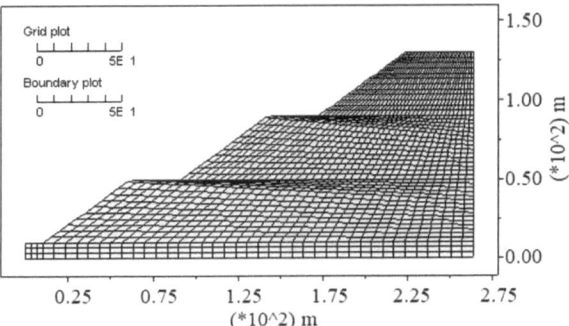

In the present study, stability analysis of the external dump containing different proportions of mix were conducted under different geometrical configurations. The overall dump height of 120.0 m was analysed under different sets of benches i.e., four benches of 30.0 m (height) each, three benches of 40.0 m (height) each, and two benches of 60.0 m (height) each. Further, the slope angle was varied from 28° at 2° interval to suggest the optimum slope angle for each set. Considering FOS value of 1.30 [20, 29], the optimum slope angle for each set of benches was suggested. The FOS of 1.30 is adequate for the recent fill and for mine dump stability [29]. Figures 3.3, 3.4 and 3.5 shows the numerically simulated geometry of external OBD dump at an overall height of 120.0 m and 28° slope angle at different bench sets.

3.5 Multiple Linear Regression (MLR) Model

MLR is a statistical analysis which builds the relationship between input and target variables. It develops a mathematical expression among dependent (input) variables and two or more (multiple) target variables. The general mathematical equation for the multiple linear regression is expressed in the following form:

Fig. 3.5 Numerically simulated OBD dump having two benches of individual height and bench slope of 60.0 m and 32° respectively

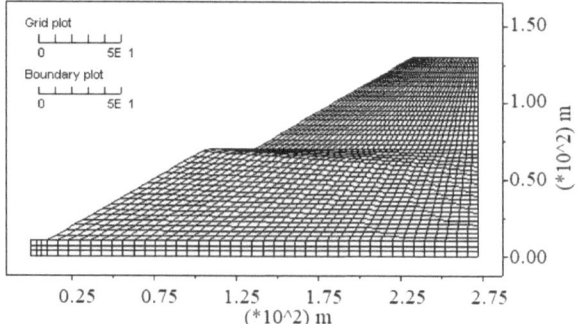

$$Y = a_1 + b_1 x_1 + b_2 x_2 + b_3 x_3 + \cdots + b_n x_n + \epsilon \qquad (3.3)$$

where,

Y	dependent variable
$x_1, x_2, x_3 \ldots x_n$	independent variable
$b_1, b_2, b_3, b_4 \ldots b_n$	regression coefficient
a_1	constant
ϵ	error

Coefficient of determination (R^2) is used to suggest the correlation among input and output variables in the model fit curve. Prediction of FOS for slope stability using MLR analysis has been successfully reported in various literatures [35–37]. MLR model was established to determine the FOS using different input variables. The developed equation represents the relationship between various input variables i.e. unit weight (γ, kN/m^3), angle of internal friction (ϕ, degree), cohesion (c, kN/m^2), number of bench (B_N), height of individual bench (H_B, m), slope angle of individual bench (β_B, degree), and one output variable i.e., FOS (F).

3.6 Artificial Neural Network (ANN) Model

ANN, a computational model, is used to train the physical structure using the information processing system. By learning, capturing, and generalising the neural network, ANN is a powerful tool for modelling the direct relation of underlying datasets. The input layer, hidden layer, and output layer are the three layers that make up an ANN architecture. Although the input layer neurons do not perform any computational operations, they do receive input from the outside environment. The hidden layer receives information from the input layer, does the calculation, and then transfers the results to the output layer. The user received the system output through the output neurons [38]. In order to reduce network error, the weights are altered and the network error is acknowledged backwards from the output layer to the input layer.

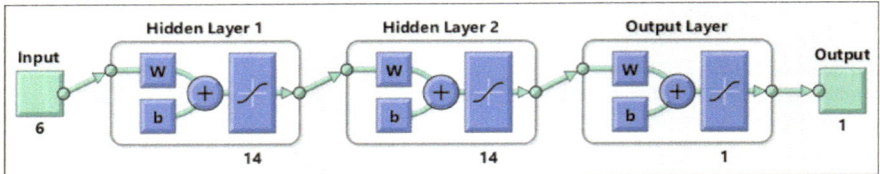

Fig. 3.6 Neural network architecture of ANN model 6-14-14-1 (N9)

Architecture of ANN Structure

The ANN model is examined using Matlab R2015a. A feed-forward back propagation network algorithm with a Levenberg–Marquardt (LM) training function is utilised to train the algorithm in this paper. When compared to other network algorithms, it gives the most helpful and appropriate training algorithm [39].

The use of ANN in several civil engineering fields has been mentioned in the literature [40–42]. The prediction of FOS of slope stability problems using ANN has been explored effectively [37, 39, 43–48].

In this study, six input parameters namely unit weight (γ, kN/m^3), angle of internal friction (ϕ, degree), cohesion (c, kN/m^2), number of bench (B_N), height of individual bench (H_B, m), and slope angle of individual bench (β_B, degree) were simulated in the input layer. FOS obtained from numerical modelling was simulated in the output layer. Architecture of the ANN model is presented in Fig. 3.6. Various ANN models have been constructed by altering the number of hidden layers (single and double layer) and the number of neurons in the hidden layer. The activation functions (AF) for the input and output layers were tansig and purelin, respectively. The hidden layer's (single and double layer) number of neurons is varied from 10 to 20 neurons in steps of 2 neurons in order to examine various ANN models. The specifics of different ANN models' architectures are displayed in Table 3.3.

ANN models from various architectures are trained to achieve a pre-set performance objective of 1e−06 across 10,000 epochs. To calculate the FOS, 476 sets at various geometrical and geomaterial parameter circumstances were numerically modelled. The entire data set is separated into two parts: in the first part, 381 data sets (about. 80%) were used to create an ANN model, and in the second part, 95 data sets (approx. 20%) were used to validate the generated model. However, in ANN, the 381 data sets are further separated into three unique groups: 70% for training, 15% for testing, and 15% for model validation [49]. The Mean Square Error (MSE) value of each architecture was calculated using Eq. (3.4).

$$MSE = \frac{1}{N} \sum_{i=1}^{N} (F_{target} - F_{predicted})^2 \tag{3.4}$$

When a model achieves the performance target with the least amount of error, it was determined to be the best neural network. Model 6-14-14-1's neural network design is seen in Fig. 3.6. The N9 model's notation 6-14-14-1 indicates that there

Table 3.3 Architecture detail of different ANN model

Model name	Architecture	Number of neurons			
		Input layer	1st hidden layer	2nd hidden layer	Output layer
N1	6-10-1	6	10	Layer absent	1
N2	6-12-1	6	12	Layer absent	1
N3	6-14-1	6	14	Layer absent	1
N4	6-16-1	6	16	Layer absent	1
N5	6-18-1	6	18	Layer absent	1
N6	6-20-1	6	20	Layer absent	1
N7	6-10-10-1	6	10	10	1
N8	6-12-12-1	6	12	12	1
N9	6-14-14-1	6	14	14	1
N10	6-16-16-1	6	16	16	1
N11	6-18-18-1	6	18	18	1
N12	6-20-20-1	6	20	20	1

are 6 neurons in the input layer, 14 neurons in each of the first and second hidden layers, and 1 neuron in the output layer.

3.7 Results and Discussions

The present study focuses to utilize fly ash in mine OBD material with and without GGBS. Specific gravity of the OBD material, fly ash, and GGBS is found to be 2.66, 2.10, and 2.83 respectively. The higher iron content of GGBS and OBD material than fly ash results in the higher specific gravity [50]. Table 3.4 presents the results of the EDX test carried out to determine the elemental composition of the OBD material, fly ash, and GGBS. The test revealed that silica (Si) is the most abundant element in both the coal mine OBD material and fly ash. On the other hand, calcium (Ca) is the most abundant element in GGBS. The mineral structure of the OBD material, fly ash, and GGBS is illustrated in Fig. 3.7. It can be observed from Fig. 3.7, that fly ash has spherical shape whereas the OBD material and GGBS show some angular structure of irregular shape. The particle size distribution curve of the OBD material and fly ash is shown in Fig. 3.8. The OBD material is classified as well-graded sand (SW) by the Unified Soil Classification System (USCS), while fly ash is classified as inorganic silt (ML). The liquid limits of the OBD material and fly ash were calculated to be 23.60% and 36.30%, respectively. Fly ash with a higher percentage of finer particles than the OBD material has a higher liquid limit. Both materials are revealed to be non-plastic. The fact that the OBD material and fly ash exhibit a negligible shrinkage limit makes them appropriate materials for various geotechnical applications [51].

Table 3.4 Chemical characterization of the materials

Elements	Elemental composition (by % weight) of materials		
	OBD material	Fly ash	GGBS
Aluminium (Al)	26.01	24.09	12.14
Silicon (Si)	56.47	63.00	29.59
Potassium (K)	03.46	02.61	00.63
Calcium (Ca)	00.64	01.12	44.80
Titanium (Ti)	02.61	02.43	00.89
Iron (Fe)	10.82	06.75	11.51

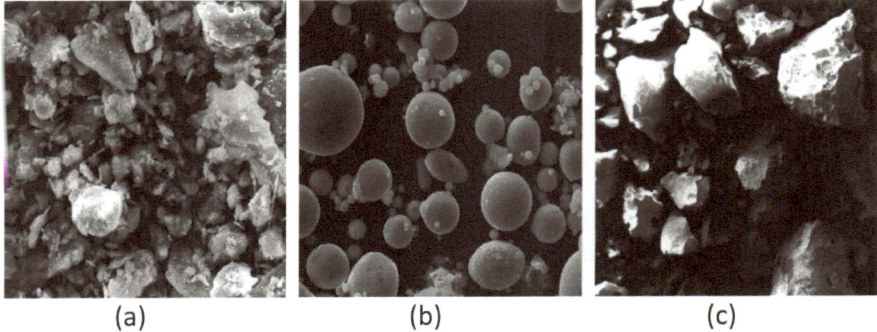

(a)   (b)   (c)

Fig. 3.7 SEM image at 5000X magnification **a** OBD material **b** fly ash **c** GGBS

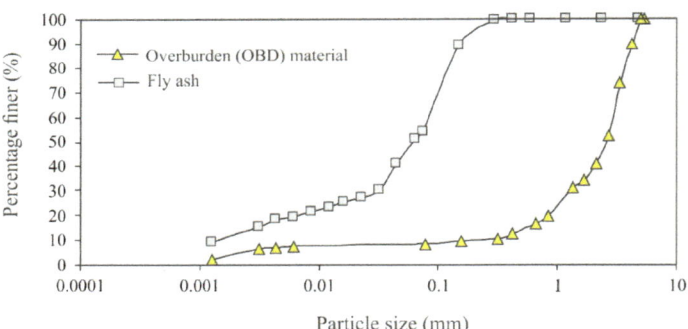

Fig. 3.8 Particle size distribution curve of OBD material and fly ash

Compaction curve of the OBD material and fly ash at standard Proctor compaction test and modified Proctor compaction test are shown in Fig. 3.9a and b respectively. MDD and OMC for the OBD material at standard Proctor test was 17.20 kN/m^3 and 14.50% respectively, and at modified Proctor test it was 19.20 kN/m^3 and 11.28% respectively. However, in case of fly ash, MDD and OMC at standard Proctor test

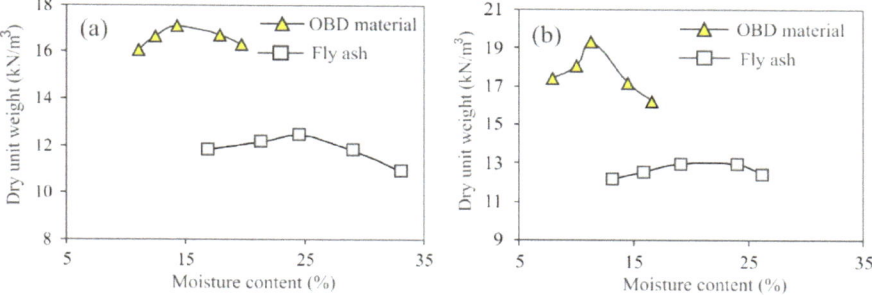

Fig. 3.9 Compaction curve of OBD material and fly ash at **a** Standard proctor compaction test **b** Modified proctor compaction test

was 12.45 kN/m^3 and 24.56% respectively, and at modified Proctor test it was 13.00 kN/m^3 and 21.50% respectively. OBD material has higher MDD than fly ash because of its higher specific gravity. However, OMC of the fly ash is higher than the OBD material.

3.7.1 Compaction Characteristics

All mixture proportions are tested with modified Proctor compaction tests (Table 3.2), and the compaction curve is shown in Fig. 3.10a and d. Table 3.5 displays the obtained MDD and OMC for all mix proportions. The findings indicate that as the proportion of fly ash in the fly ash-OBD mixture increases, the MDD decreases while the OMC increases. This decrease in MDD can be attributed to the reduced specific gravity of fly ash. According to Behera and Mishra [16], the non-cohesive characteristic of fly ash diminishes the MDD of a fly ash-OBD combination. Furthermore, the addition of GGBS to a fly ash-OBD mixture induces an increase in MDD due to its greater specific gravity [52]. The reduction in MDD with the incorporation of fly ash and GGBS results in the reduction of stress induced due to self-weight. Thus, it can be advantageous for the improvement of stability of coal mine dump.

3.7.2 Unconsolidated Undrained (UU) Triaxial Test

The UU test is a popular laboratory experiment for determining strength factors such as angle of internal friction and cohesion. Table 3.6 shows the assessed shear strength characteristics for all of the mix proportions that were tested. It was discovered that the cohesiveness value of a fly ash-OBD mix increases up to 20% fly ash content and thereafter declines. However, when the fly ash content in the fly ash-OBD combination increases, the angle of internal friction

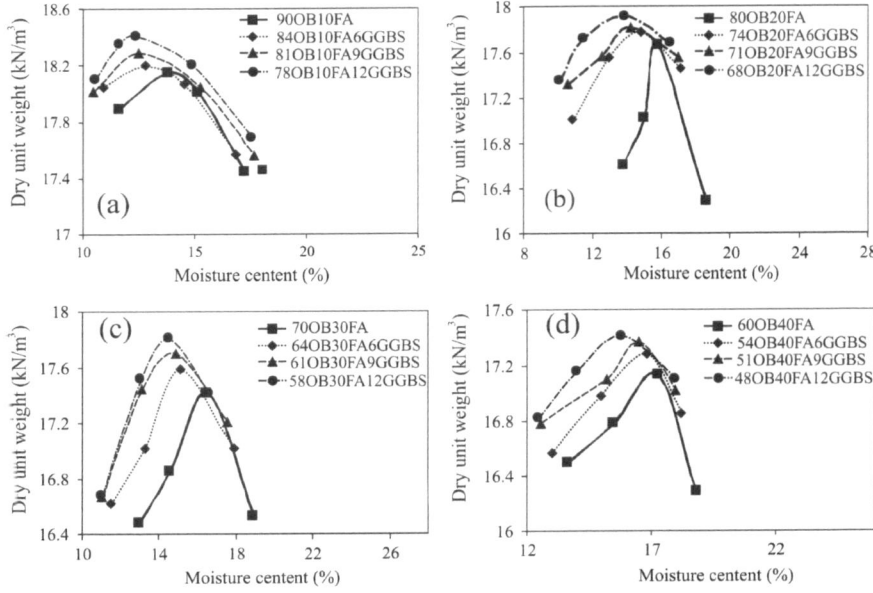

Fig. 3.10 Compaction curve of OBD material at different proportion of GGBS and fly ash **a** 10% fly ash **b** 20% fly ash **c** 30% fly ash **d** 40% fly ash

	Material name	MDD (kN/m³)	OMC (%)
Table 3.5 MDD and OMC of different samples	90OBD10F	18.20	13.20
	84OBD10F6GGBS	18.28	12.90
	81OBD10F9GGBS	18.35	12.40
	78OBD10F12GGBS	18.45	12.20
	80OBD20F	17.56	15.95
	74OBD20F6GGBS	17.75	14.75
	71OBD20F9GGBS	17.86	14.35
	68OBD20F12GGBS	17.92	13.75
	70OBD30F	17.35	16.40
	64OBD30F6GGBS	17.48	15.20
	61OBD30F9GGBS	17.72	14.80
	58OBD30F12GGBS	17.86	14.30
	60OBD40F	17.10	17.20
	54OBD40F6GGBS	17.35	16.70
	51OBD40F9GGBS	17.42	16.60
	48OBD40F12GGBS	17.47	15.87

Table 3.6 Shear strength parameter of all mix composites

Material name	Cohesion (kPa)	Angle of internal friction
100OBD	20.46	31.10°
90OBD10F	28.05	29.26°
84OBD10F6GGBS	29.00	29.16°
81OBD10F9GGBS	32.26	28.16°
78OBD10F12GGBS	36.68	28.10°
80OBD20F	29.25	27.58°
74OBD20F6GGBS	31.16	27.02°
71OBD20F9GGBS	34.53	26.13°
68OBD20F12GGBS	37.12	26.31°
70OBD30F	28.05	27.18°
64OBD30F6GGBS	29.15	27.13°
61OBD30F9GGBS	32.28	26.33°
58OBD30F12GGBS	36.12	26.28°
60OBD40F	27.85	25.20°
54OBD40F6GGBS	29.28	25.12°
51OBD40F9GGBS	31.75	24.48°
48OBD40F12GGBS	35.56	24.10°

decreases. It might be because of the spherical form of the fly ash (Fig. 3.7). Furthermore, the inclusion of GGBS enhances the cohesiveness value of the fly ash-OBD mixture. The increase in cohesion value might be attributed to the filling of cavities in the OBD material by finer fly ash and GGBS particles [52], as well as the cementitious character of GGBS particles [23, 24]. Cohesion value was increased by 81.42% for 68% OB material, 20% fly ash, and 12% GGBS as compared to 100% OB material.

3.7.3 Stability Analysis

Stability analysis of external dump containing different proportions of OBD material, fly ash, and GGBS under different geometrical configuration has been performed using FLAC/Slope software. The overall dump height of 120.0 m was analysed under different sets of benches i.e., four benches of 30.0 m (height) each, three benches of 40.0 m each, and two benches of 60.0 m each. Further, the slope angle was varied from 28° at 2° interval. The failure pattern of the external dump containing 78% OBD, 10% fly ash, and 12% GGBS at an overall height of 120.0 m and different bench set are presented in Figs. 3.11, 3.12 and 3.13.

Fig. 3.11 Failure pattern, shear strain rate contour, and FOS value plot for the dump containing 78% OBD, 10% fly ash, and 12% GGBS at four benches with individual height and slope angle of 30.0 m and 44° respectively

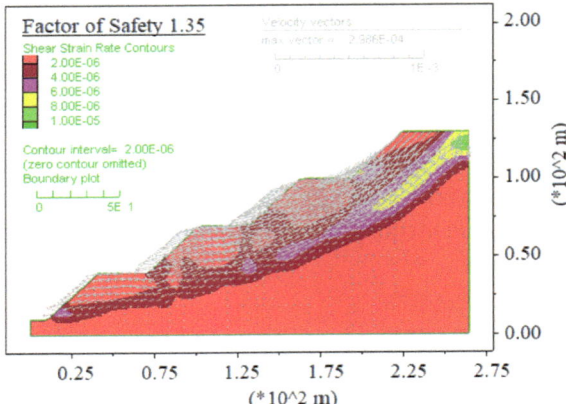

Fig. 3.12 Failure pattern, shear strain rate contour, and FOS value plot for the dump containing 78% OBD, 10% fly ash, and 12% GGBS at three benches with individual height and slope angle of 40.0 m and 38° respectively

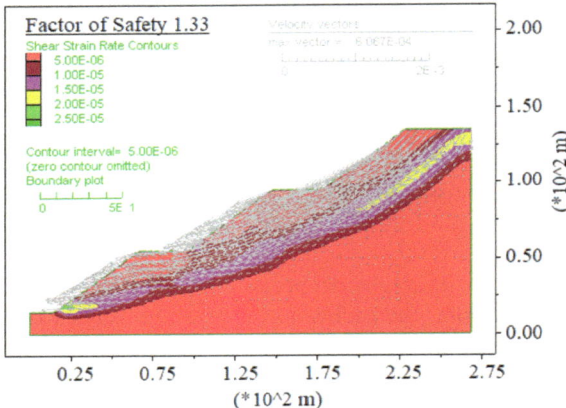

Fig. 3.13 Failure pattern, shear strain rate contour, and FOS value plot for the dump containing 78% OBD, 10% fly ash, and 12% GGBS at two benches with individual height and slope angle of 60.0 m and 32° respectively

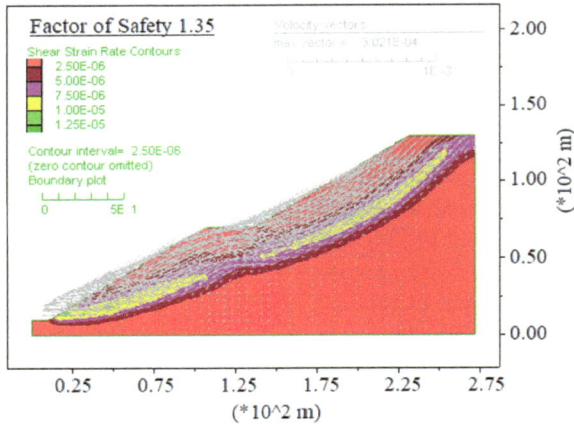

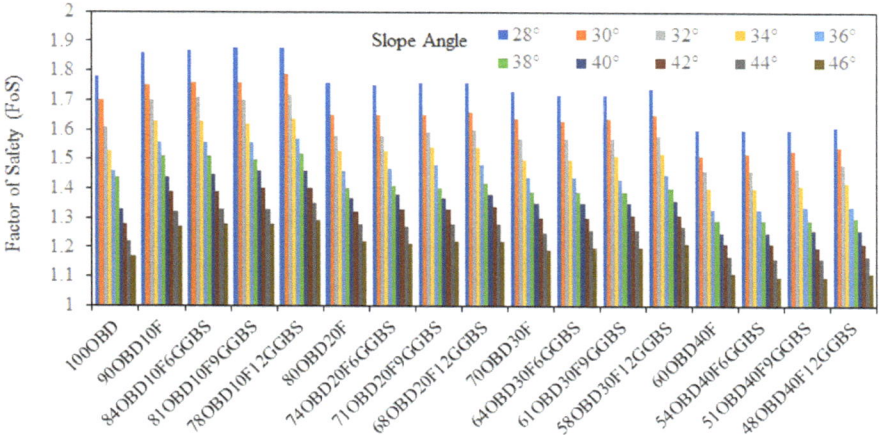

Fig. 3.14 Variation of FOS with various slope angles and different samples at four benches of 30.0 m individual height

The variation of FOS of all the mix proportions at different slope angles and bench sets are presented in Figs. 3.14, 3.15 and 3.16. Based on the FOS value of 1.30 [20], the suitable slope angle for each sample under different bench sets has been reported in Table 3.7. It can be found from Figs. 3.14, 3.15 and 3.16 that with increase of the slope angle, the FOS decreases significantly. With increase in the no of benches, the FOS of the external dump increases but it may reduce the dump capacity. The maximum dump capacity can be achieved with the two benches of individual height and slope angle of 60.0 m and 32° respectively compared to other dump geometry for the OBD material containing 10% fly ash with and without GGBS.

3.7.4 Multiple Linear Regression (MLR) Model

In the present work, MLR model was established to predict the FOS of external dump under different geometrical and geomaterial properties. The established model enables the relationship amongst explanatory (input) variables i.e., unit weight (γ, kN/m^3), angle of internal friction (ϕ, degree), cohesion (c, kN/m^2), number of bench (BN), height of individual bench (H_B, m), slope angle of the individual bench (β_B, degree), and one response (output) variable i.e., FOS (F). In numerical modelling, a total of 476 sets of different geometrical and geomaterial parameters were analysed to determine the FOS. In which around 80% data sets (i.e., 381 data sets) were used to build the MLR model and the remaining 20% data sets (95 data sets) were used for the validation of the developed model. The MLR model for the prediction of FOS is presented in Eq. 3.5.

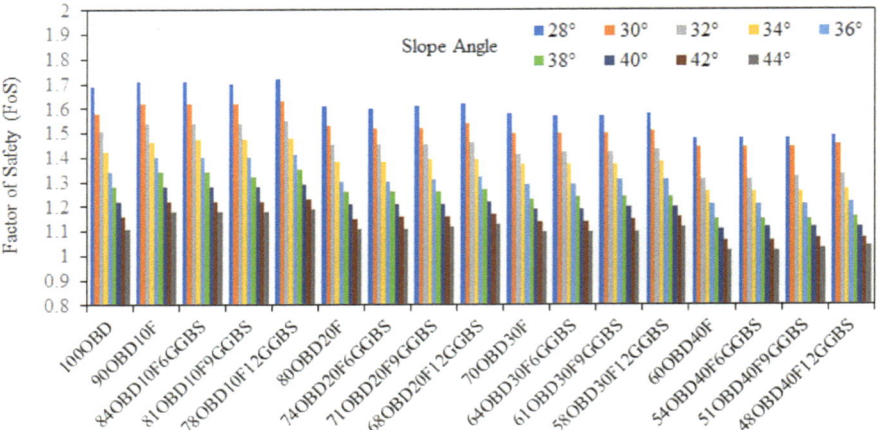

Fig. 3.15 Variation of FOS with various slope angles and different samples at three benches of 40.0 m individual height

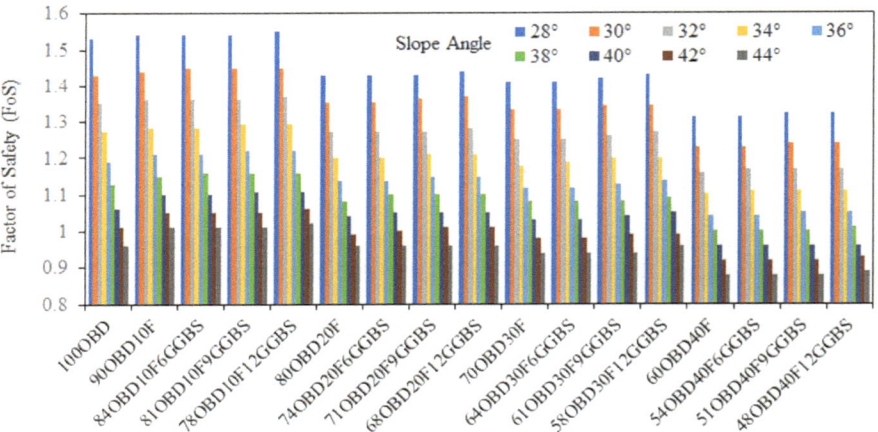

Fig. 3.16 Variation of FOS with various slope angles and different samples at two benches of 60.0 m individual height

$$FoS = 0.971 - 0.033\gamma + 0.011c + 0.050\phi - 0.030\beta_B - 0.002H_B + 0.132B_N$$
$$(3.5)$$

Regression statistics of MLR model is presented in Table 3.8. Coefficients of input parameters and their corresponding Probability value (P-value) are presented in Table 3.9. The significance of the developed MLR model can be explained by coefficient of determination (R^2) and P value. It can be observed from Tables 3.8 and

Table 3.7 Suggested individual slope angle at various bench sets for different samples

Mix proportions/ Material name	Suggested bench slope angle (β_B) of at various bench sets		
	β_B at four benches of 30.0 m individual height	β_B at three benches of 40.0 m individual height	β_B at two benches of 60.0 m individual height
100OBD	40°	36°	32°
90OBD10F	44°	38°	32°
84OBD10F6GGBS	44°	38°	32°
81OBD10F9GGBS	44°	38°	32°
78OBD10F12GGBS	44°	38°	32°
80OBD20F	42°	36°	30°
74OBD20F6GGBS	42°	36°	30°
71OBD20F9GGBS	42°	36°	30°
68OBD20F12GGBS	42°	36°	30°
70OBD30F	42°	34°	30°
64OBD30F6GGBS	42°	34°	30°
61OBD30F9GGBS	42°	36°	30°
58OBD30F12GGBS	42°	36°	30°
60OBD40F	36°	32°	28°
54OBD40F6GGBS	36°	32°	28°
51OBD40F9GGBS	36°	32°	28°
48OBD40F12GGBS	36°	32°	28°

3.9 that developed MLR is statistically significant because of the higher coefficient of determination value i.e., $R^2 = 0.98$ and very small p-value (< 0.05) of each parameter.

Validation of MLR Model

Validation of the developed MLR model has been carried out by comparing the FOS value obtained from numerical modelling with FOS predicted by the MLR model. 20% of the total data set (i.e., 95 data sets), which was not used for the development of the MLR model, has been used to compare the observed and predicted values of FOS. The comparison between observed and predicted FOS is presented in Fig. 3.17

Table 3.8 Regression statistics of developed MLR model

Regression statistics	
Multiple R	0.99
R^2	0.98
Adjusted R^2	0.98
Standard error	0.02
Observations	381

Table 3.9 Corresponding coefficients and P-value of the various parameters

Parameter	Coefficients	Standard error	t-statistics	P-value
Intercept	0.970	0.084896	11.4348	3.60E−26
γ, kN/m^3	−0.033	0.005949	−5.54004	5.71E−08
c, kN/m^2	0.011	0.000422	25.59937	3.11E−84
ϕ, degree	0.050	0.001797	27.95101	1.28E−93
B_N	−0.030	0.000259	−115.672	7.3E−295
H_B, m	−0.002	0.000594	−2.93665	0.003523
β_B, degree	0.132	0.009033	14.60046	1.59E−38

Fig. 3.17 Comparison of FOS value for 95 data sets (20%) between observed (numerical model) and predicted (MLR model)

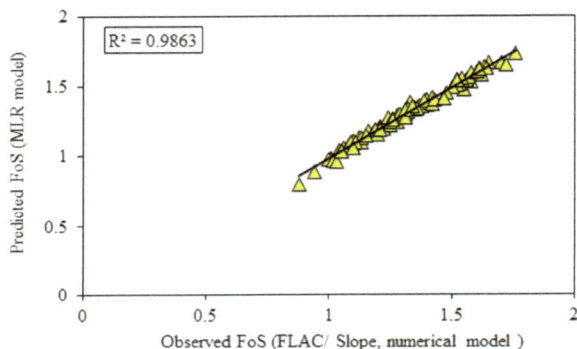

which indicates close agreement between observed and predicted FOS with high value of coefficient of determination value i.e., $R^2 = 0.9863$.

3.7.5 ANN Model

Different architectures of ANN model has been developed by varying the number of hidden layer and number of neurons to ascertain the best model. Matlab R2015a has been used to train the different models. The MSE value and obtained coefficient of determination (R2) of testing, training, and validation for all the different ANN models are presented in Table 3.10.

Table 3.10 shows that, despite having identical R^2 values, the ANN model (N9) with architecture 6-14-14-1 is the best model compared to others since it first achieves the performance objective of 1e−06 with a low MSE value. However, as the number of neurons in the hidden layer increases, like in the case of N10, N11, and N12, the performance target is attained with a smaller epoch value.

Table 3.10 MSE and coefficient of determination value of different ANN model

Model name	Architecture	Coefficient of determination (R^2)			MSE value
		Testing	Training	Validation	
N1	6-10-1	0.998	0.997	0.999	0.0000808
N2	6-12-1	0.999	0.997	0.999	0.0001575
N3	6-14-1	0.999	0.984	0.999	0.0001079
N4	6-16-1	0.999	0.999	0.999	0.0001380
N5	6-18-1	0.999	0.992	0.998	0.0001784
N6	6-20-1	0.998	0.998	0.998	0.0001753
N7	6-10-10-1	0.998	0.998	0.999	0.0001687
N8	6-12-12-1	0.997	0.999	0.999	0.0002189
N9	6-14-14-1	0.999	0.993	0.998	0.0001349
N10	6-16-16-1	0.998	0.998	0.998	0.0001700
N11	6-18-18-1	0.998	0.997	0.998	0.0002429
N12	6-20-20-1	0.999	0.996	0.997	0.0002686

Validation of ANN Model

The ANN model was validated by simulating the remaining 20% of data sets (95 data sets) that were not used while creating the ANN model. The projected FOS (for 20% of the data sets) from the ANN model (N9) was compared to the observed FOS from numerical modelling. Figure 3.18 shows the FOS value of the ANN model and the numerical model (FLAC/Slope) for 95 data sets. Figure 3.19 shows the coefficient of determination value (R^2) for validation of the ANN model (N9) and comparison of observed and projected FOS. Figures 3.18 and 3.19 show a high level of statistical dependability ($R^2 = 0.9968$) between the ANN model (N9) and the numerical model.

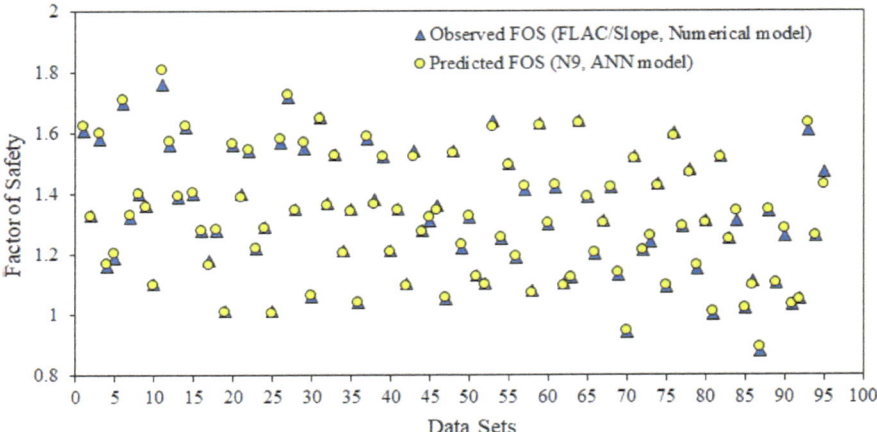

Fig. 3.18 FOS value of ANN model and numerical model (FLAC/Slope) for 95 data sets

Fig. 3.19 Comparison between observed (numerical model) and predicted FOS (N9, ANN model)

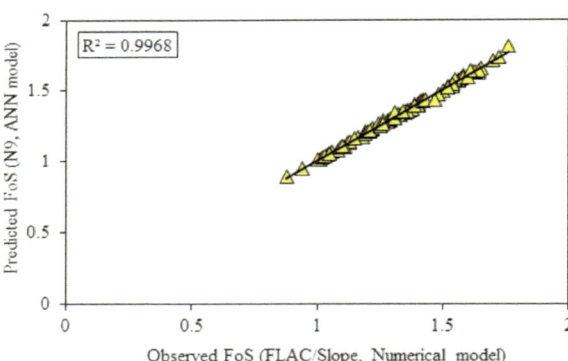

3.8   Conclusion

Following conclusions are drawn from the present study:

1. The MDD decreased and OMC increased with the addition of fly ash in the OBD material. Further addition of GGBS in the fly ash-OBD material, results in increase in MDD due to the higher specific gravity of GGBS compared to fly ash and OBD material.
2. The cohesion value of the OBD material increased with the addition of fly ash. However, a significant decrease in the angle of internal friction was observed with increase in the fly ash content. Maximum improvement in cohesion value

of fly ash-OBD mix was observed for mix proportions containing 80% OBD material and 20% fly ash.

3. Addition of GGBS in the fly ash-OBD mix shows a significant result in strength improvement. The cohesion value increases and the angle of internal friction decreases with the addition of GGBS content in the fly ash-OBD material. Maximum cohesion value was observed for mix proportions containing 68% OBD material, 20% fly ash, and 12% GGBS. Cohesion value was increased by 81.42% for 68% OB material, 20% fly ash, and 12% GGBS as compared to 100% OB material.

4. Based on the numerical modelling of the external dump of 120.0 m overall height, suitable slope angle has been suggested at different bench combinations. However, maximum dump capacity can be achieved with the two benches of 60.0 m (height) each at 32° slope angle for the OBD material containing 10% fly ash with and without GGBS.

5. The developed MLR model and ANN model (N9) with architecture 6-14-14-1 predict the FOS at high accuracy with coefficient of determination value of 0.9863 and 0.9968 respectively.

6. Based on the results findings, it can be recommended to incorporate 10% to 20% fly ash with 12% GGBS with mine OB material for the improvement of mine dump and for providing a suitable sub-base material for mine haul road.

References

1. Richards, B. G., Coulthard, M. A., & Toh, C. T. (1981). Analysis of slope stability at Goonyella Mine. *Canadian Geotechnical Journal, 18*(2), 179–194.
2. Dawson, R. F., Morgenstern, N. R., & Stokes, A. W. (1998). Liquefaction flowslides in Rocky Mountain coal mine waste dumps. *Canadian Geotechnical Journal, 35*(2), 328–343.
3. Poulsen, B., Khanal, M., Rao, A. M., Adhikary, D., & Balusu, R. (2014). Mine overburden dump failure: A case study. *Geotechnical and Geological Engineering, 32*(2), 297–309.
4. Upadhyay, O. P., Sharma, D. K., Singh, D. P. (1990). Factors affecting stability of waste dumps in mines. *International Journal of Surface Mining, Reclamation and Environment, 4*(3), 95–99.
5. Koner, R., & Chakravarty, D. (2016). Characterisation of overburden dump materials: A case study from the Wardha valley coal field. *Bulletin of Engineering Geology and the Environment, 75*(3), 1311–1323.
6. Edil, T. B., Sandstrom, L. K., & Berthouex, P. M. (1992). Interaction of inorganic leachate with compacted pozzolanic fly ash. *Journal of Geotechnical Engineering, 118*(9), 1410–1430.
7. Kumar, B. R. P., & Sharma, R. S. (2004). Effect of Fly Ash on Engineering Properties of Expansive Soils. *Journal of Geotechnical Geoenvironmental Engineering, 130*(July), 764–767.
8. Prabakar, J., Dendorkar, N., & Morchhale, R. K. (2004). Influence of fly ash on strength behavior of typical soils. *Construction and Building Materials, 18*(4), 263–267.
9. Ghosh, A., & Subbarao, C. (2006). Tensile strength bearing ratio and slake durability of class F fly ash stabilized with lime and gypsum. *Journal of Materials in Civil Engineering, 18*(1), 18–27.
10. Punthutaecha, K., Puppala, A. J., Vanapalli, S. K., & Inyang, H. (2006). Volume change behaviors of expansive soils stabilized with recycled ashes and fibers. *Journal of Materials in Civil Engineering, 18*(2), 295–306.

11. Sharma, N. K., Swain, S. K., & Sahoo, U. C. (2012). Stabilization of a clayey soil with fly ash and lime: A micro level investigation. *Geotechnical and Geological Engineering, 30*(5), 1197–1205.
12. Pandian, N. S. (2004). Fly ash characterization with reference to geotechnical applications. *Journal of the Indian Institute of Science, 84,* 189–216.
13. Rajak, T. K., Yadu, L., & Pal, S. K. (2019). Analysis of slope stability of fly ash stabilized soil slope. In I.V. Anirudhan & V.B. Maji (Eds.), *Geotechnical Applications,* Lecture Notes in Civil Engineering (vol. 13, no. 51, pp. 119–126).
14. Rajak, T. K., Yadu, L., Chouksey, S. K., & Pal, S. K. (2017). Strength characteristics of fly ash stabilized soil embankment and stability analysis using numerical modelling. In *Indian Geotechnical Conference 2017 GeoNEst.*
_5. Tannant, D. D., & Kumar, V. (2000). Properties of fly ash stabilized haul road construction materials. *International Journal of Surface Mining, Reclamation Environment, 14*(2), 121–135.
16. Behera, B., & Mishra, M. K. (2012). California bearing ratio and Brazilian tensile strength of mine overburden-fly ash-lime mixtures for mine haul road construction. *Geotechnical and Geological Engineering, 30*(2), 449–459.
17. Behera, B. (2016). Geotechnical properties of lime stabilized fly-ash mine overburden for haul road. *International Journal of Advances in Mechanical and Civil Engineering, 3*(6), 7–12.
18. Mahamaya, M., & Das, S. K. (2016). Characterization of mine overburden and fly ash as a stabilized pavement material. *Particulate Science and Technology, 35*(6), 660–666.
19. Mallick, S. R., & Mishra, M. K. (2017). Evaluation of clinker stabilized fly ash-mine overburden mix as sub-base construction material for mine haul roads. *Geotechnical and Geological Engineering, 35*(4), 1629–1644.
20. Rajak, T. K., Yadu, L., Chouksey, S. K., & Dewangan, P. K. (2018). Stability analysis of mine overburden dump stabilized with fly ash. *International Journal of Geotechnical Engineering,* 1–11.
21 Cockrell, C. F., & Leonard, J. W. (1970). *Characterization and utilization studies of limestone modified fly ash.* Coal Research Bureau Report 60, West Virginia University, Morgantown.
22. Behera, B., & Mishra, M. K. (2012). Strength behaviour of surface coal mine overburden–fly ash mixes stabilised with quick lime. *International Journal of Mining, Reclamation and Environment, 26*(1), 38–54.
23. Wild, S., Kinuthia, J. M., Jones, G. I., & Higgins, D. D. (1998). Effects of partial substitution of lime with ground granulated blast furnace slag (GGBS) on the strength properties of lime-stabilised sulphate-bearing clay soils. *Engineering, 51,* 37–53.
24. Rai, A., Prabakar, J., Raju, C. B., & Morchalle, R. K. (2002). Metallurgical slag as a component in blended cement. *Construction and Building Materials, 16*(8), 489–494.
25. Al-Rawas, A. A. (2002). Microfabric and mineralogical studies on the stabilization of an expansive soil using cement by-pass dust and some types of slags. *Canadian Geotechnical Journal, 39*(5), 1150–1167.
26. Cokca, E., Yazici, V., & Ozaydin, V. (2009). Stabilization of expansive clays using granulated blast furnace slag (GBFS) and GBFS-Cement. *Geotechnical and Geological Engineering, 27*(4), 489–499.
27. Yadu, L., & Tripathi, R. K. (2013). Effects of granulated blast furnace slag in the engineering behaviour of stabilized soft soil. *Procedia Engineering, 51,* 125–131.
28. Sharma, A. K., & Sivapullaiah, P. V. (2016). Ground granulated blast furnace slag amended fly ash as an expansive soil stabilizer. *Soils and Foundations, 56*(2), 205–212.
29. Pradhan, S. P., Vishal, V., Singh, T. N., & Singh, V. K. (2014). Optimisation of dump slope geometry vis-à-vis fly ash utilisation using numerical simulation. *American Journal of Mining and Metallurgy, 2*(1), 1–7.
30. ASTM D854-14. (2014). Standard test methods for specific gravity of soil solids by water pycnometer. In *ASTM International.*
31. ASTM-D6913M. (2017). Standard test methods for particle-size distribution (Gradation) of soils using sieve analysis. In *ASTM International.*

32. ASTM D698-12. (2012). ASTM-D698-12: Standard test methods for laboratory compaction characteristics of soil using standard effort. In *ASTM International*.
33. ASTM D1557 (2012) Standard test methods for laboratory compaction characteristics of soil using modified effort. In *ASTM International* (pp. 1–14).
34. ASTM D2850. (2015). Standard test method for unconsolidated-undrained triaxial compression test on cohesive soils. In *ASTM International*.
35. Kostic, S., Vasoviv, N., & Sunaric, D. (2016). Slope stability analysis based on experimental design. *International Journal of Geomechanics, 16*(1), 1–11.
36. Yellishetty, M., & Darlington, W. J. (2011). Effects of monsoonal rainfall on waste dump stability and respective geo-environmental issues: A case study. *Environment and Earth Science, 63*(6), 1169–1177.
37. Chakraborty, A., & Goswami, D. (2017). Prediction of slope stability using multiple linear regression (MLR) and artificial neural network (ANN). *Arabian Journal of Geosciences, 10*(17), 1–11.
38. Guo, Z., & Uhrig, R. E. (1992). Use of artificial neural networks to analyze nuclear power plant performance. *Nuclear Technology, 99*(1), 36–42.
39. Gupta, M. M., Jin, L., & Homma, N. (2003). *Static and dynamic neural networks*. Wiley & Sons Inc.
40. Goh, A. T. C., Wong, K. S., & Broms, B. B. (1995). Estimation of lateral wall movements in braced excavations using neural networks. *Canadian Geotechnical Journal, 32*(6), 1059–1064.
41. Flood, I., & Kartam, N. (1994). Neural networks in civil engineering I: Principles and understanding. *Journal of Computing in Civil Engineering, 8*(2), 131–148.
42. Cheng, J., Li, Q. S., & Cheng Xiao, R. (2008). A new artificial neural network-based response surface method for structural reliability analysis. *Probabilistic Engineering Mechanics, 23*(1), 51–63.
43. Ni, S. H., Lu, P. C., & Juang, C. H. (1996). A fuzzy neural network approach to evaluation of slope failure potential. *Microcomputers in Civil Engineering, 11*(1), 59–66.
44. Sakellariou, M. G., & Ferentinou, M. D. (2005). A study of slope stability prediction using neural networks. *Geotechnical and Geological Engineering, 23*(4), 419–445.
45. Das, S. K., Biswal, R. K., Sivakugan, N., & Das, B. (2011). Classification of slopes and prediction of factor of safety using differential evolution neural networks. *Environment and Earth Science, 64*(1), 201–210.
46. Verma, A. K., Singh, T. N., Chauhan, N. K., & Sarkar, K. (2016). A hybrid FEM–ANN approach for slope instability prediction. *Journal of The Institute of Engineers: Series A, 97*(3), 171–180.
47. Choobbasti, A. J., Farrokhzad, F., & Barari, A. (2009). Prediction of slope stability using artificial neural network (Case study: Noabad, Mazandaran, Iran). *Arabian Journal of Geosciences, 2*(4), 311–319.
48. Abdalla, J. A., Attom, M. F., & Hawileh, R. (2015). Prediction of minimum factor of safety against slope failure in clayey soils using artificial neural network. *Environment and Earth Science, 73*(9), 5463–5477.
49. Ghaboussi, J., Sidarta, D. E., & Lade, P. V. (1994). Neural network based modelling in geomechanics. In *Computer Methods and Advances in Geomechanics* (pp. 153–164).
50. Kumar, H., & Mishra, M. K. (2014). Optimization and evaluation of fly ash composite properties for geotechnical application. *Arabian Journal of Geosciences, 8*(6), 3713–3726.
51. Mallick, S. R., & Mishra, M. K. (2013). Geotechnical characterization of clinker-stabilized fly ash—Coal mine overburden mixes for subbase of mine haul road. *Coal Combustion and Gasification Products, 5*(1), 49–56.
52. Akinmusuru, J. O. (1991). Potential beneficial uses of steel slag wastes for civil engineering purposes. *Resources, Conservation and Recycling, 5*(1), 73–80.

Chapter 4
Performance of Coal Mine Overburden Dump Slope Under Earthquakes Using Extended Finite Element Method Based Voronoi Tessellation Scheme

Madhumita Mohanty⊙, Rajib Sarkar⊙, and Sarat Kumar Das⊙

4.1 Introduction

In India, the production of coal by the removal of the waste material overlying the coal seams is majorly being performed using opencast mining. The sloping structures formed due to the piling up of these heterogeneous wastes are termed as coal mine overburden (OB) dump slopes [1]. The OB dumps have to be accommodated in the limited available space, which leads to rise in their heights, thus making them vulnerable to failures. There has been an increase in the number of accidents due to the OB dump failures [2]. Various problems occur on account of the presence of OB dumps: precious land is lost and existing ecosystems are endangered [3]; occurrence of environmental hazards [4]; several houses and people are buried [5] and mining activities get interrupted [6].

Earthquake is one of the most important factors that significantly influences the OB dump failures. Although uncontrollable seismic energy is produced, the Indian guidelines do not specify any regulations considering the seismic effects. Further, evaluation of the slope stability of the heterogeneous OB dump is a major challenge.

In the present study, the Voronoi tessellation scheme has been used to represent the innumerable discontinuities of the heterogeneous OB dump. The presence of numerous discontinuities make the use of finite element method (FEM) troublesome

M. Mohanty · R. Sarkar (✉) · S. K. Das
Indian Institute of Technology (Indian School of Mines), Dhanbad, Jharkhand, India
e-mail: rajib@iitism.ac.in

M. Mohanty
e-mail: madhumita.iitism@gmail.com

S. K. Das
e-mail: saratdas@iitism.ac.in

as meshing becomes difficult in the problem domain. Thus, extended finite element method (XFEM) has been used here as it overcomes the difficulties faced while using FEM. The commercial software package, RS2 v11.013 2021 [7] is used to couple the XFEM and Voronoi tessellation scheme, then the dynamic analyses were carried out. A set of ten earthquakes were chosen [8], which covered a wide range of strong ground motion characteristics. Ultimately, the amplification ratios [9] were estimated at four key points for investigating the seismic performance of the heterogeneous OB dump as well as a similar homogeneous OB dump. On account of negligible literature being available considering the seismic damage of OB dump slopes, the current study would be helpful in the preparation of design guidelines for OB dumps in earthquake prone areas.

4.2 Background of XFEM

FEM is advantageous when the domain of the problem is moderately jointed, whereas in case of a heavily jointed domain, the process of meshing becomes tough. The utilization of XFEM overcomes the abovementioned problem. Several studies have successfully implemented XFEM [10] as it considers the mesh and joints to be independent of each other [11–13]. The effect of joints is considered in an implicit manner when they cross the element [14]. Firstly, the position of the joints are disregarded and the discretization of the domain of the problem occurs independently. Then, addition of enriched nodes occur to all those elements which are intersected by joints. The term enriched node is used to denote the nodes of an element that have been intersected by a joint. Additional degrees of freedom (DOFs) are provided to each node according to the number of joints present in the element. A domain consisting of two joints has been discretized and is shown in Fig. 4.1.

The coordinates of a point in the domain is represented as (x, y) and defined as per the local coordinates of the joint. The Heaviside function for a joint has been illustrated in Fig. 4.2. Heaviside function, $H(x)$, is used to consider the discontinuity in the element and is expressed as follows [12]:

$$H(x) = \begin{cases} +1 & y > 0 \\ -1 & y < 0 \end{cases} \tag{4.1}$$

Next, the displacement, $u(x)$ is expressed as follows [10]:

$$u(x) = \sum_{i \in I} N_i(x)\overline{u_i} + \sum_{j \in J} N_j(x)(H(x) - H(x_j))\widehat{u_j} \tag{4.2}$$

where, N_i denotes the shape function for the ith node, I denotes the set of all nodes in the domain, J is the set of enriched nodes, $\overline{u_i}$ is the set of standard DOFs and $\widehat{u_j}$ is

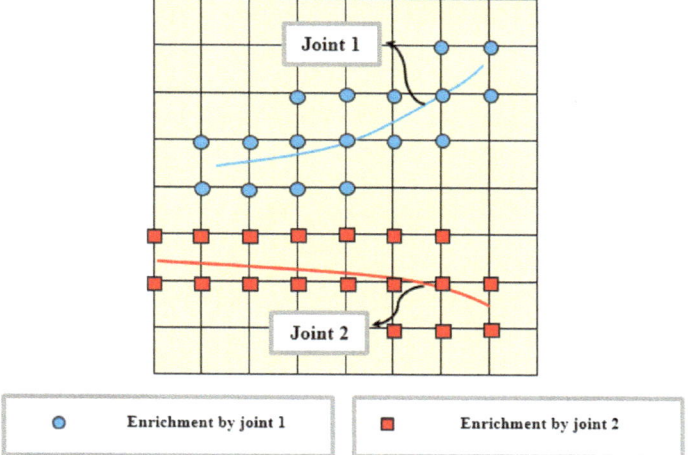

Fig. 4.1 Discretization of a problem domain consisting of two joints (adopted from Moallemi et al. [14])

Fig. 4.2 Heaviside function for a joint (adopted from Moës et al. [12])

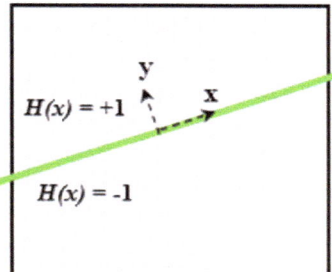

the set of enriched DOFs. The expression, $H(x) - H(x_j)$ represents the enrichment function across the joint.

In the Voronoi tessellation scheme, random shapes and sizes of Voronoi blocks can be represented efficiently. The interface existing between the Voronoi blocks are called as Voronoi contacts (or joints). The size of the particles in the OB dump range from less than 0.0001 m and may exceed 1 m [1]. To specify the density of the Voronoi contacts, an average length of 1 m of the edges of Voronoi polygons has been used in this study, so that the length of a joint in the problem domain may be lesser, greater or equal to 1 m. To obtain the most irregular shaped network of Voronoi polygons, the regularity of the polygon shape was considered to be "irregular", and was thus provided as input.

4.3   Generation of the OB Dump Model

The properties considered for the Voronoi tessellation scheme are based on multi-channel analysis of surface waves (MASW) test performed on the OB dump of Jambad open cast coal mine (India) and utilizing several correlations [15–18]. The properties of the Voronoi blocks and Voronoi contacts have been obtained from Mohanty et al. [19] and it also contains the detailed procedure of evaluating them [19]. The material properties have been summarized in Table 4.1. Mohr–Coulomb model is most applicable for general engineering studies, general rock mechanics and very well represents the joints in rock. It can be appreciably applied to loose rock mass. Thus, in the present study, Mohr–Coulomb constitutive model was assigned to the Voronoi blocks and Mohr–Coulomb slip criterion was assigned to the Voronoi contacts.

In the present study, the geometrical features of the OB dump model have been decided according to the regulations stated in Coal Mines Regulations 2017 [20]. The schematic diagram for the heterogeneous OB dump has been given in Fig. 4.3. The boundary conditions as well as the four key points have been shown in it. The average Voronoi joint length has been considered to be 1 m. In order to prevent the reflections of the input wave, absorbing and transmitting boundary conditions have been provided at the bottom and the lateral boundary. Rest of the surfaces were considered as free surfaces. Both the horizontal and vertical movements were restricted at the bottom boundary of the OB dump slope model. The movement along the horizontal direction was restricted at its lateral boundary. V_x and V_y are the velocities along the horizontal and vertical directions respectively. The maximum size of the mesh element used was less than one-tenth of the wavelength associated with the highest frequency of the input wave as per Kuhlemeyer and Lysmer [21]. Three percentage of Rayleigh damping was used. Finally, the set of ten earthquakes were applied individually [8] as input motions. Firstly, the input motion was applied as velocity history at the base of the OB dump model. Considering the absorbing boundary condition, the velocity record had to be converted to a stress record by means of the compliant base condition. Then, the input wave propagates upward. Lastly, amplification ratios were estimated for the four key points to indicate the

Table 4.1 Properties of the Voronoi tessellation scheme used in heterogeneous OB dump [19]

Voronoi blocks	Property	Unit weight	Elastic modulus	Poisson's ratio	Cohesion	Friction angle	Tensile strength
	Unit	(kN/m^3)	(kPa)	–	(MPa)	(°)	(MPa)
	Value	14.5	304,850	0.35	0	33.32	0
Voronoi contacts	Property	Normal stiffness	Shear stiffness	Cohesion of joint	Friction angle of joint		
	Unit	(kPa/m)	(kPa/m)	(MPa)	(°)		
	Value	430,000	70,000	0	30.02		

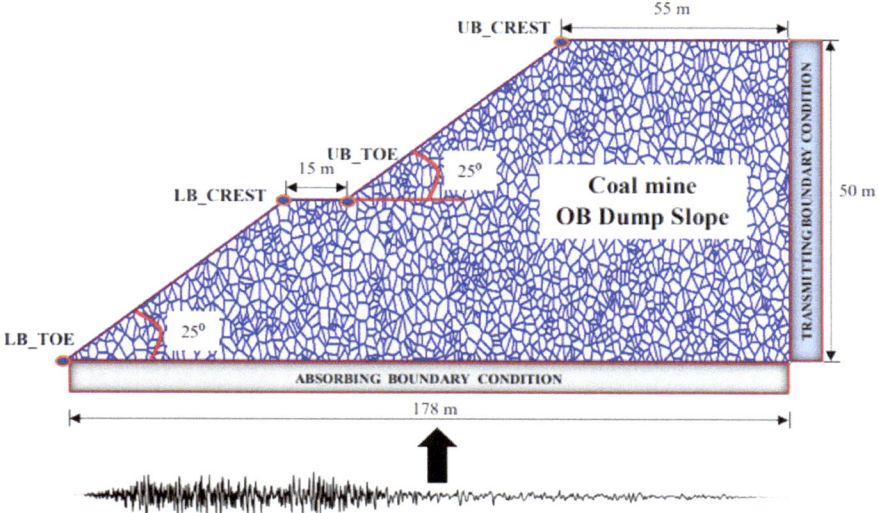

Fig. 4.3 Schematic diagram of the XFEM coupled Voronoi tessellated heterogeneous OB dump model

seismic damage. Further, a comparative study was also done between heterogeneous and homogeneous OB dump slopes. The homogeneous OB dump slope was prepared using similar geometrical features, boundary conditions and material properties of Voronoi block as the heterogeneous OB dump, but it was devoid of the Voronoi contacts.

4.4   Results and Discussions

A set of ten earthquake input motions [8] varying in their strong ground motion parameters were considered for the dynamic analyses. Amplification ratio (ratio between the peak ground acceleration (PGA) at the key point of the OB dump to the PGA of the applied earthquake input motion) was used to quantify the damage due to earthquake. As already discussed in the earlier sections, XFEM was coupled with the Voronoi tessellation scheme to prepare the model of heterogeneous OB dump. Thereafter, a homogeneous model was made considering identical geometrical features, boundary conditions and material properties. Next, another step was taken to compare the seismic performance of heterogeneous and homogeneous OB dumps. The upcoming sections elucidate the results in a detailed manner.

Table 4.2 Details of the amplification ratios at the toe of lower bench of the OB dump slope for the earthquake motions considered in the present study [8]

Earthquake	PGA (g)	Amplification ratio	
		Heterogeneous OB dump	Homogeneous OB dump
Chi Chi (1999)	0.18	1.04	1.45
Coyote (1979)	0.12	1.00	1.00
Imperial valley (1979)	0.17	1.03	1.02
Kobe (1995)	0.82	1.07	1.14
Kocaeli (1999)	0.22	0.93	1.19
Loma Gilroy (1989)	0.17	0.98	1.42
Mammoth lake (1980)	0.43	1.06	1.11
Northridge (1994)	0.22	1.38	1.66
Parkfield (2004)	0.36	1.05	0.94
Whittier narrows (1987)	0.19	0.93	0.88

4.4.1 Amplification Ratio at the Toe of the Lower Bench of the OB Dump Slope

The details of the PGA values of the ten input motions, and the amplification ratios thus obtained considering the toe of the lower bench of OB dump are given in Table 4.2. The pictorial representation showing the variation of amplification ratio for the earthquake motions considered in the present study for the toe of the lower bench of the OB dump slope has been provided in Fig. 4.4. The amplification ratios observed did not proportionately increase with the increase in PGA values. It was further observed that in some cases the amplification ratio was higher for the homogeneous OB dump while in other cases it was higher for the heterogeneous OB dump. The values fail to show any regular trend. The reason may be attributed to the influence of the other strong ground motion parameters like Arias intensity, bracketed duration and predominant frequency.

4.4.2 Amplification Ratio at the Crest of the Lower Bench of the OB Dump Slope

Similar to the earlier section, Table 4.3 was prepared considering the crest of the lower bench. The graphical representation of the Table 4.3 is provided in Fig. 4.5. In case of both the OB dumps, the amplification ratios did not show increasing pattern along with the increase in PGA values of the earthquakes. It is worthwhile to mention here that the heterogeneous OB dump showed higher amplification ratios in comparison to the homogeneous OB dumps. The crest is the highest point on the lower bench, thus it is more vulnerable to movement during earthquakes. This might have been

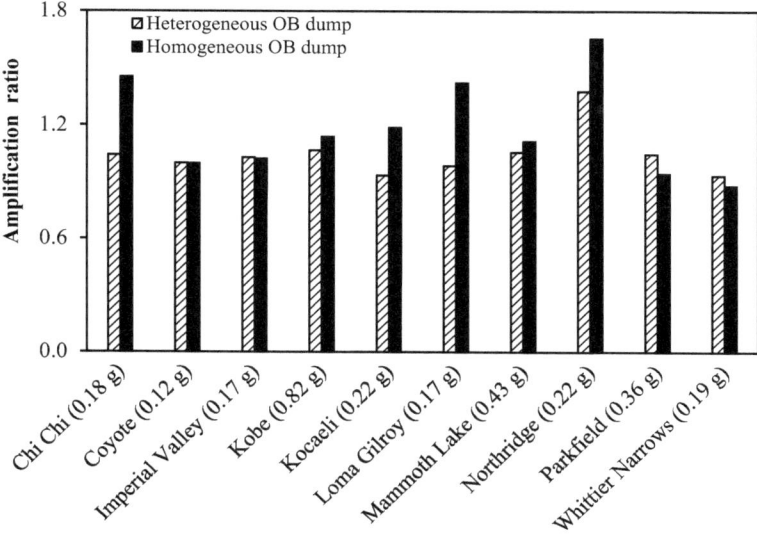

Fig. 4.4 Variation of amplification ratio at the toe of lower bench of the OB dump slope for the earthquake motions considered in the present study

due to the sliding and opening of the nodes at the ends of the Voronoi contacts which are present in numerous amount in the heterogeneous OB dump, whereas in the homogeneous OB dump, there is absolute absence of Voronoi contacts. Therefore, during the progress of the earthquakes, there might have been considerable movement of the Voronoi blocks which would have led to higher amplification ratio in case of the heterogeneous OB dumps.

4.4.3 Amplification Ratio at the Toe of the Upper Bench of the OB Dump Slope

Table 4.4 was prepared along lines similar to the previous two tables considering the toe of the upper bench. The changes in the amplification ratios were elucidated digrammatically in Fig. 4.6. Even here, the amplification ratios are higher for the heterogeneous OB dumps as compared to the homogeneous OB dumps. Moreover, any increasing pattern was not noticed for the amplification ratios along with the increase in PGA values.

The inferences drawn here are almost similar to the earlier section as the crest of the lower bench and the toe of the upper bench are present on the same bench.

Table 4.3 Details of the amplification ratios at the crest of lower bench of the OB dump slope for the earthquake motions considered in the present study [8]

Earthquake	PGA (g)	Amplification ratio	
		Heterogeneous OB dump	Homogeneous OB dump
Chi Chi (1999)	0.18	1.01	0.84
Coyote (1979)	0.12	1.24	0.80
Imperial valley (1979)	0.17	1.76	1.02
Kobe (1995)	0.82	0.96	0.70
Kocaeli (1999)	0.22	1.51	1.04
Loma Gilroy (1989)	0.17	1.67	1.07
Mammoth lake (1980)	0.43	1.38	0.72
Northridge (1994)	0.22	1.68	1.03
Parkfield (2004)	0.36	1.35	0.75
Whittier narrows (1987)	0.19	2.05	1.10

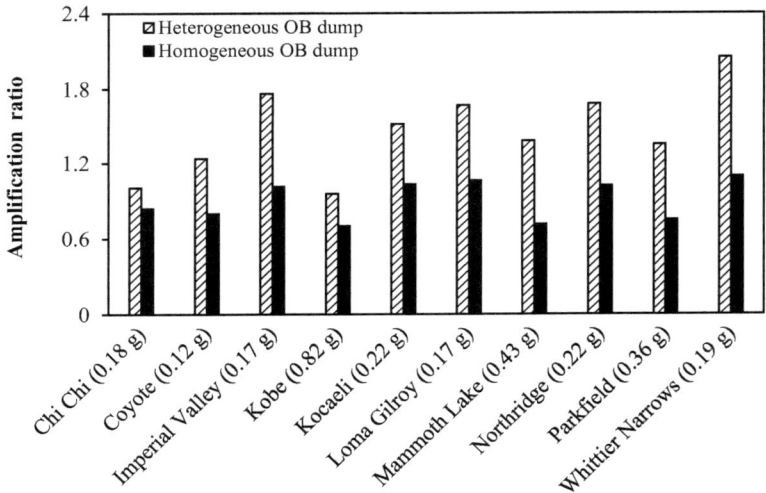

Earthquake (PGA g) LB_CREST

Fig. 4.5 Variation of amplification ratio at the crest of lower bench of the OB dump slope for the earthquake motions considered in the present study

4.4.4 Amplification Ratio at the Crest of the Upper Bench of the OB Dump Slope

The PGA of the considered earthquakes and the resulting amplification ratios for both types of OB dumps considering the crest of the upper bench have been summarized

Table 4.4 Details of the amplification ratios at the toe of upper bench of the OB dump slope for the earthquake motions considered in the present study [8]

Earthquake	PGA (g)	Amplification ratio	
		Heterogeneous OB dump	Homogeneous OB dump
Chi Chi (1999)	0.18	0.99	0.81
Coyote (1979)	0.12	1.19	0.85
Imperial valley (1979)	0.17	1.51	1.02
Kobe (1995)	0.82	1.02	0.76
Kocaeli (1999)	0.22	1.36	1.02
Loma Gilroy (1989)	0.17	1.68	1.08
Mammoth lake (1980)	0.43	1.18	0.66
Northridge (1994)	0.22	1.51	1.10
Parkfield (2004)	0.36	1.29	0.85
Whittier narrows (1987)	0.19	1.53	0.91

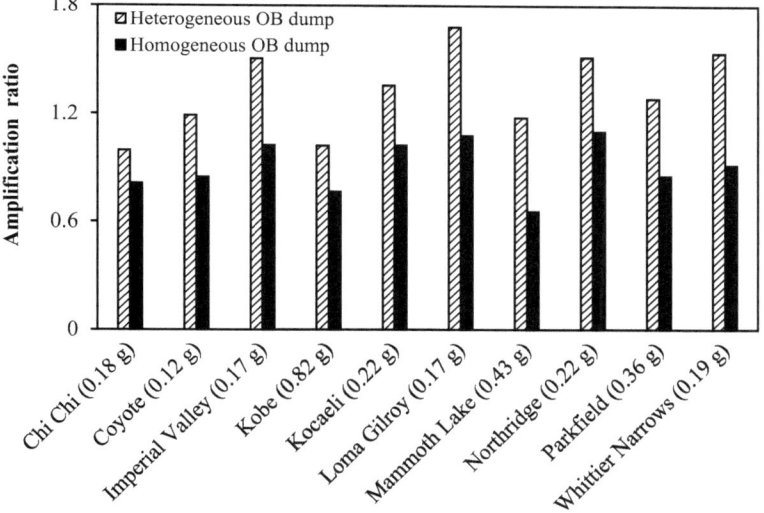

Earthquake (PGA g) UB_TOE

Fig. 4.6 Variation of amplification ratio at the toe of upper bench of the OB dump slope for the earthquake motions considered in the present study

Table 4.5 Details of the amplification ratios at the crest of upper bench of the OB dump slope for the earthquake motions considered in the present study [8]

Earthquake	PGA (g)	Amplification ratio	
		Heterogeneous OB dump	Homogeneous OB dump
Chi Chi (1999)	0.18	0.14	0.81
Coyote (1979)	0.12	0.15	0.83
Imperial valley (1979)	0.17	0.23	0.82
Kobe (1995)	0.82	0.24	0.76
Kocaeli (1999)	0.22	0.27	0.93
Loma Gilroy (1989)	0.17	0.32	1.11
Mammoth lake (1980)	0.43	0.16	0.60
Northridge (1994)	0.22	0.30	0.99
Parkfield (2004)	0.36	0.30	0.70
Whittier narrows (1987)	0.19	0.22	0.87

in Table 4.5 and its pictorial representation is given in Fig. 4.7. The values of the amplification ratios did not follow any regular pattern. Moreover, it was seen that the amplification ratios for the heterogeneous OB dumps were relatively lesser than the corresponding ones for the homogeneous OB dumps. The reason behind this may be the lesser continuity of the system near the crest of the upper bench for the propagation of wave and the higher damping induced during the earthquakes. The discontinuity of the materials in the heterogeneous OB dump slope model may have hindered the propagation of the wave to the top. The dynamic response of the heterogeneous OB dump slope is also dependent on other factors like material properties of the Voronoi blocks and properties of the Voronoi contacts, while in the homogeneous OB dump, the Voronoi contacts are absent.

4.5 Conclusions

The heterogeneous OB dump model was formed by coupling XFEM with Voronoi tessellation scheme and its seismic performance was studied using ten different earthquakes. The amplification ratio could be properly evaluated considering heterogeneity. The amplification ratios did not follow any regular pattern based on the PGA values of the earthquakes. At the crest of the lower bench and toe of the upper bench, the amplification ratios were higher for the heterogeneous OB dump, whereas at the crest of the upper bench the amplification ratios were quite insignificant for the said OB dump. The evaluation of the seismic performance of the heavily jointed OB dump could be successfully performed by incorporating the Voronoi tessellation scheme in XFEM.

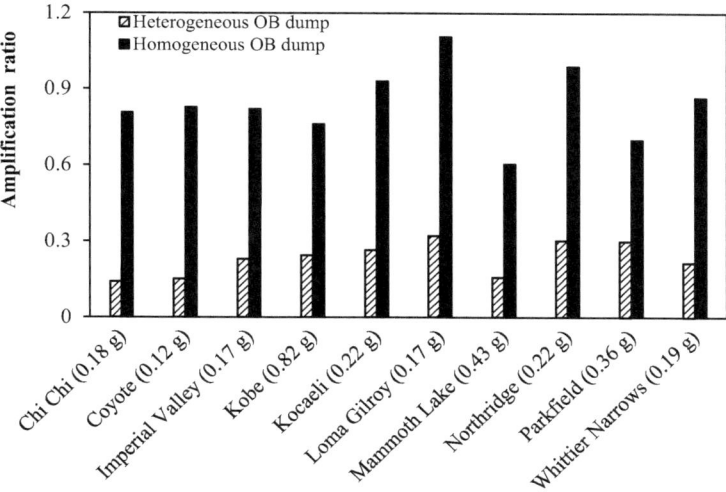

Earthquake (PGA g) UB_CREST

Fig. 4.7 Variation of amplification ratio at the crest of upper bench of the OB dump slope for the earthquake motions considered in the present study

References

1. Nayak, P. K., Dash, A., & Dewangan, P. (2020). Design considerations for waste dumps in Indian opencast coal mines—A critical appraisal. In *Proceedings of 2nd International Conference on Opencast Mining Technology and Sustainability* (pp. 19–31).
2. Directorate General of Mines Safety (DGMS). (2010). *Design, control and monitoring of pit and dump slopes in opencast mines*. Director General of Mines Safety, Circular no. 02, Dhanbad, India.
3. Tripathi, N., Singh, R. S., & Chaulya, S. K. (2012). Dump stability and soil fertility of a coal mine spoil in Indian dry tropical environment: A long-term study. *Environmental Management, 50*(4), 695–706.
4. Adibee, N., Osanloo, M., & Rahmanpour, M. (2013). Adverse effects of coal mine waste dumps on the environment and their management. *Environmental Earth Sciences, 70,* 1581–1592.
5. Duc, D. M., Hieu, N. M., Sassa, K., Hamasaki, E., Khang, D. Q., Miyagi, T. (2014). Analysis of a deep-seated landslide in the Phan Me coal mining dump site, Thai Nguyen Province, Vietnam. Proceedings of World Landslide Forum 3, 2–6 June 2014, Beijing, 1. In K. Sassa, P. Canuti. & Y. Yin (Eds.), *Landslide science for a safer geoenvironment* (pp. 373–377). Springer.
6. Wang, J., & Chen, C. (2017). Stability analysis of slope at a disused waste dump by two-wedge model. *International Journal of Mining, Reclamation and Environment, 31*(8), 575–588.
7. RS2 v11.013. (2021). *2D finite element based software*. Rocscience Inc, Toronto, Ontario, Canada.
8. Pacific Earthquake Engineering Research Center (PEER) Ground Motion Database Web Application, PEER. http://peer.berkeley.edu/smcat/ (2010).
9. Bottari, C., Albano, M., Capizzi, P., Alessandro, A. D., Doumaz, F., Martorana, R., Moro, M., & Saroli, M. (2018). Recognition of earthquake-induced damage in the Abakainon Necropolis (NE Sicily): Results from geomorphological, geophysical and numerical analyses. *Pure and Applied Geophysics, 175,* 133–148.

10. Fries, T. P., & Belytschko, T. (2010). The extended/generalized finite element method: An overview of the method and its applications. *International Journal for Numerical Methods in Engineering, 84*, 253–304.
11. Belytschko, T., & Black, T. (1999). Elastic crack growth in finite elements with minimal remeshing. *International Journal for Numerical Methods in Engineering, 45*(5), 601–620.
12. Moës, N., Dolbow, J., & Belytschko, T. (1999). A finite element method for crack growth without remeshing. *International Journal for Numerical Methods in Engineering, 46*(1), 131–150.
13. Moës, N., & Belytschko, T. (2002). Extended finite element method for cohesive crack growth. *Engineering Fracture Mechanics, 69*(7), 813–833.
14. Moallemi, S., Curran, J. H., & Yacoub, T. (2018). On modeling rock slope stability problems using XFEM. In *Paper presented at the 2nd International Discrete Fracture Network Engineering Conference*, Seattle, Washington, USA (pp. 1–9).
15. Anbazhagan, P., Uday, A., Moustafa, S. S. R., & Al-Arifi, N. S. N. (2016). Correlation of densities with shear wave velocities and SPT N values. *Journal of Geophysics and Engineering, 13*, 320–341.
16. Barton, N. (1976). The shear strength of rock and rock joints. *International Journal of Rock Mechanics and Mining Sciences and Geomechanics Abstracts, 13*(9), 255–279.
17. Kumar, R., Bhargava, K., & Choudhury, D. (2016). Estimation of engineering properties of soils from field SPT using random number generation. *INAE Letters, 1*, 77–84.
18. Tan, X., Zhao, M., Zhu, Z., & Jin, Y. (2019). Elastic properties calibration approach for discrete element method model based on Voronoi tessellation method. *Geotechnical and Geological Engineering, 37*(3), 2227–2236.
19. Mohanty, M., Sarkar, R., & Das, S. K. (2022). In-situ investigation on coal mine overburden dump slope and its seismic stability considering heterogeneity. *European Journal of Environmental and Civil Engineering*, 1–25.
20. Directorate General of Mines Safety (DGMS). (2017). Coal Mines Regulations, Notification, New Delhi, dated 27.11.2017. Ministry of Labor and Employment, Directorate General of Mines Safety.
21. Kuhlemeyer, R. L., & Lysmer, J. (1973). Finite element method accuracy for wave propagation problems. *Journal of the Soil Mechanics and Foundations Division, 99*, 421–427.

Chapter 5
Slope Stability Analysis of Coalmine Overburden Dump Using a Probabilistic Approach

Ashutosh Kumar⑩, **Sarat Kumar Das**⑩, **Lohitkumar Nainegali**⑩, and **Krishna R. Reddy**⑩

5.1 Introduction

Coal mining operations involve the excavation of coal deposits from the earth's crust [1]. During this process, large amounts of soil and rock, known as overburden (OB), are removed, and deposited in the vicinity of the mining areas. These areas are commonly known as overburden dumps or spoil piles.

Overburden dumps are typically created by dumping the waste materials from coal mining operations in a designated area. The dumping process results in the formation of a large mound, which can range from several meters to hundreds of meters in height. Overburden dumps can cover large areas and are often visible from a distance due to their size and elevation. A typical overburden dump is shown in Fig. 5.1. The generation of overburden dumps has several adverse effects on the environmental such as environmental pollution including air, soil, water contamination, and land degradation. Another major concern associated with overburden dumps is the risk of slope failure. Dumping of large amounts of overburden can result in the destabilization of the soil, which can cause damage to nearby structures, equipment,

A. Kumar (✉) · S. K. Das · L. Nainegali
Indian Institute of Technology (Indian School of Mines), Dhanbad, Jharkhand, India
e-mail: ashu.18dr0044@cve.iitism.ac.in

S. K. Das
e-mail: saratdas@iitism.ac.in

L. Nainegali
e-mail: lohitkumar@iitism.ac.in

K. R. Reddy
University of Illinois Chicago, Chicago, IL, US
e-mail: kreddy@uic.edu

© The Author(s), under exclusive license to Springer Nature Singapore Pte Ltd. 2024   75
S. K. Das et al. (eds.), *Geoenvironmental and Geotechnical Issues of Coal Mine Overburden and Mine Tailings*, Springer Transactions in Civil and Environmental Engineering, https://doi.org/10.1007/978-981-99-6294-5_5

Fig. 5.1 Schematic of typical overburden dump, Jharkhand, India

loss of lives, and hinderance in mining activity [2]. Therefore, a safe geotechnical analysis is needed to ascertain the safety of the dumps.

The slope stability of these overburden dumps is critical for the safety of personnel and the surrounding environment. Further the high heterogeneity of the overburden material can significantly affect the slope stability of these dumps [3]. Heterogeneity refers to the variation in the physical and mechanical properties of the material. This variation can arise due to differences in the composition, density, moisture content, grain size, and internal structure of the material [4]. Thus, this wide variation in the material property can be handled by performing a probabilistic analysis. However, the slope stability of the coalmine overburden material was analyzed in previous studies, assuming it to be homogeneous and using constant material properties (deterministic), neglecting the possibility of spatial variability and uncertainty [5, 6].

In this study, a 2D limit equilibrium probabilistic slope stability analysis has been carried out. The dump slope used in the study consists of a 1m thick vegetative layer at the top. The results from the deterministic analysis of the vegetated slope are compared with the bare slope. Further, a coefficient of variation (CoV) in the shear strength properties of the overburden has been considered and Monte Carlo simulations have been performed with multiple realizations to assess the variations in the probability of failure (P_f) and reliability index (β).

5.2 Methodology

5.2.1 Overburden Dump Geometry

The study is performed on a full-scale coalmine OB dump representing the actual dump existing in Dhanbad, Jharkhand, India. The bare dump rests on a ground surface that consists of 90 m deep foundation strata consisting of multiple strata resembling the opencast mine lithology [7]. The multi-layered soil consists of a medium-grained sandstone reaching up to 30 m from the ground level, extended by a 10 m thick coal seam. The adjacent layer comprises of a 35 m thick coarse-medium-grained sandstone underlain by a thick coal seam of 15 m thickness. The geometry of the overburden dump (above the ground) comprises the slope (bare) with four benches (W1, W2, W3, W4) equal to 25 m at an equal height (H1, H2, H3, H4, H5) of 25 m (total height = 130m) and a slope angle (θ) equal to 2V:1H for each slope face, as shown in Fig. 5.2a. The overall slope angle of the dump was 37.5° which complied with the safety guidelines of the Directorate General of Mines Safety, India [8]. The geometry of the dump slope with vegetation was the same with the addition of an extra region of thickness equal to 1m throughout the slope face (Fig. 5.2b). The present study compares the result obtained from the vegetated dump with the results obtained by Kumar et al. [9] from the bare dump under similar conditions.

5.2.2 Slope Stability Analysis

The shear strength estimation was done by using Mohr–Coulomb failure criteria. A static general limit equilibrium approach (GLE) was employed to determine the slope factor of safety (*FoS*). The GLE method is based on the moment and force equilibrium to determine the *FoS* and it is the most rigorous method for solving circular and non-circular failure surfaces. The slope search method was selected as the search method for finding the minimum (global) *FoS* for circular slip surfaces. The PLAXIS LE CONNECT edition v21 was used for the slope stability analysis.

The *FoS* with respect to horizontal force equilibrium is as follows:

$$(F_s) = \frac{\sum[c'l\cos\alpha + (N - ul)\tan\phi'\cos\alpha]}{\sum N\sin\alpha} \tag{5.1}$$

The *FoS* with respect to moment equilibrium is as follows:

$$(F_s) = \frac{\sum[c'l\cos\alpha + (N - ul)\tan\phi'\cos\alpha)]}{\sum N\sin\alpha} \tag{5.2}$$

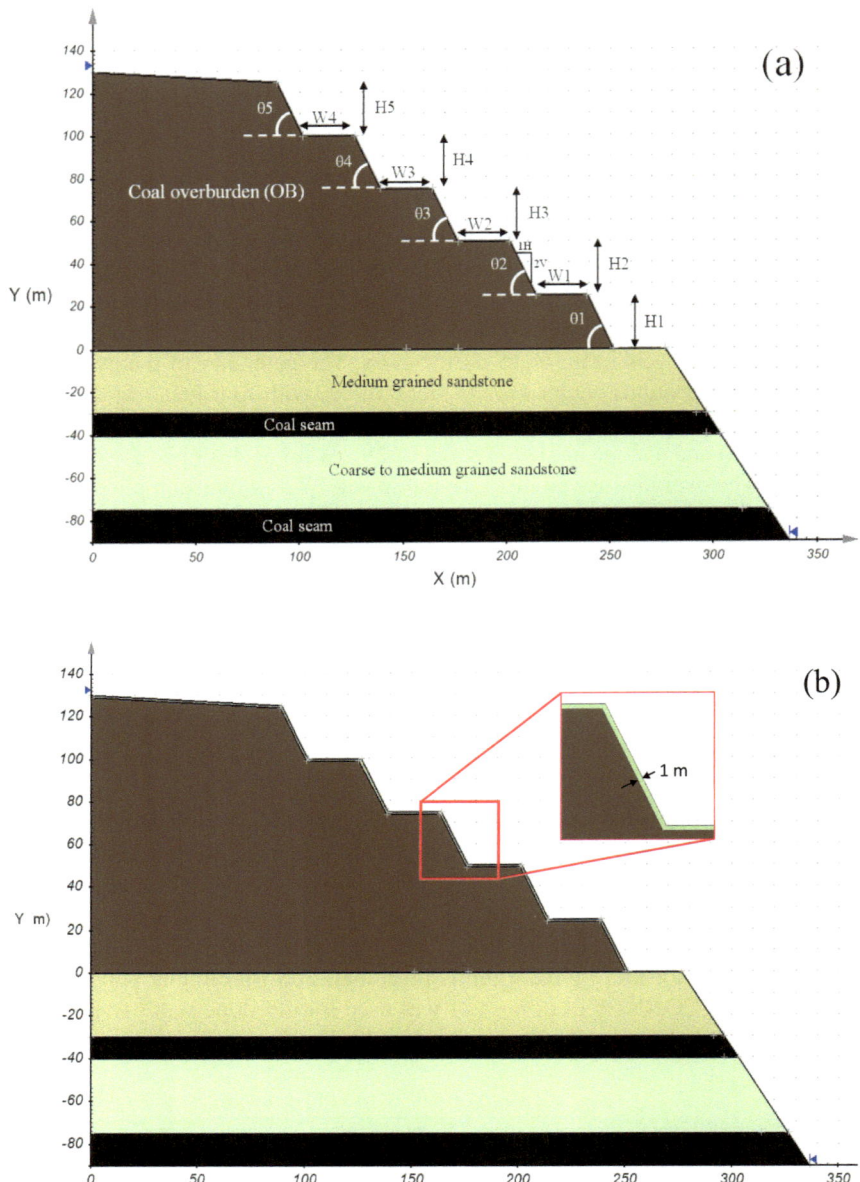

Fig. 5.2 Schematic numerical model for engineered dump **a** without vegetation (bare) and **b** with vegetation

Table 5.1 Statistical parameters of coalmine OB from the literature

Statistical parameters	Unit weight	Friction angle	Cohesion
Overburden material			
Range	14.0–20.7 kN/m^3	8°–40°	0–72 kPa
Average$_{range}$	17.35 kN/m^3	24°	36 kPa
Standard deviation (σ)	2.13	9.12	17.49
Coefficient of variation (*CoV*, %)	12.75%	33.03%	49.72%
Mean (μ)	16.73 kN/m^3	27.63°	35.13 kPa
Overburden material with roots			
Mean (μ)	16.73 kN/m^3	27.63°	80.13 kPa

where c' = effective cohesion of the material at the base of a slice; l and α are geometric properties; N = normal force on the slice of the base; u = pore-water pressure; ϕ' = effective angle of internal friction at the base of a slice.

5.2.3 Material Properties

The statistical parameters, including the mean (μ), standard deviation (σ), and coefficient of variation (CoV) in the unit weight (γ, kN/m^3), cohesion (c, kPa), and friction angle (ϕ, °) values, were calculated using published data from previous studies conducted on various coal mines worldwide [5, 6, 10–17]. This approach aimed to highlight the global spatial variability linked to the coalmine OB. The presence of root provides reinforcement to the soil, which is reflected in the form of soil cohesion, and friction angle being unaltered [18, 19]. The material properties assigned to the vegetative layer has the same value of unit weight and friction angle as that of overburden, However, the value of cohesion was increased by 45 kPa i.e., c equal to 85.13 kPa. This value was selected based on a study performed on the Sisam tree grown for the biological stabilization of the mine dumps [20]. The material properties used for the coalmine overburden and overburden with roots are presented in Table 5.1. The *CoV* (%) values were applied to the material properties under five different cases (Table 5.2) in the form of σ.

5.2.4 Probabilistic Analysis

It is possible to use reliability-based procedures to assess the uncertainties related to material properties in slope stability analysis [21]. These procedures involve treating model parameters as continuous random variables with known probability density functions and distribution parameters, which can be estimated from measured data to

Table 5.2 *CoV* in the input material properties for the probabilistic analysis

Case	Coefficient of variation (*CoV*, %)		
	Unit weight (%)	Cohesion (%)	Angle of friction (%)
1	1	10	10
2	2	20	15
3	3	30	20
4	4	40	25
5	5	50	30

obtain unbiased estimates of the sample mean and standard deviation. In this study, the log-normal distribution is used to analyze each material property of overburden. The design parameters, including friction angle, cohesion, and unit weight, are treated as random variables, and their distribution is evaluated using the probability density function (PDF). An alternative method is to use the cumulative distribution function (CDF) to determine the probability of a variable being equal to or less than a specific value. The probability of failure (P_f) can be calculated using probabilistic methods, which consider the variability of input parameters and indicate the likelihood of failure at a particular level. The log-normal PDF, ($f(x)$) and CDF, ($g(x)$) in terms of FoS are given below:

$$f(FoS) = \frac{1}{FoS\sigma\sqrt{2\pi}} exp\left(-\frac{(ln(FoS) - \mu)^2}{2\sigma^2}\right) \tag{5.3}$$

$$g(FoS) = \frac{1}{2}\left[1 + erf\left(\frac{ln(FoS - \mu)}{\sigma\sqrt{2}}\right)\right] \tag{5.4}$$

If a result is deterministic and has a higher factor of safety (*FoS*), it may provide incomplete and misleading information because the probability of failure (P_f) associated with it may also be higher. When it comes to risk-based and cost-effective design methods, the P_f (%) is typically favoured over deterministic indicators [22, 23]. It is calculated as the number of times an analysis gives a *FoS* less than one to the total number of *FoS* analyses.

The Monte Carlo Method (MCM) was employed to conduct a probability-based performance evaluation of the overburden dump by conducting 26,690 trials based on the range of coefficients of variation (*CoVs*) or standard deviation. Various groups of X input variables, represented by $x_1, x_2, ..., x_n$, were generated and evaluated using the log-normal probability density function (PDF). Each of the randomly generated sets was utilized to calculate a *FoS*(X) realization, which was then used to define the PDF of FoS(X). The statistics for the output outcomes can be presented in the probabilistic framework by assessing the reliability index (β), which is given as:

$$\beta = \frac{E[FoS] - 1}{\sigma[FoS]} \tag{5.5}$$

where, $E[FoS]$ = expected FoS; $\sigma[FoS]$ = standard deviation of FoS.

5.3 Results and Discussions

5.3.1 Deterministic Analysis

The analysis performed by Kumar et al. [9] considering a fixed value of the shear strength parameter for the bare slope yielded a $FoS = 1.08$ with the radius of the critical slip surface equal to 61.25 m. In the present study, the same dump geometry with a vegetative layer yielded a $FoS = 1.115$ having a radius of slip surface equal to 203.37 m. Figure 5.3a shows the location of the critical slip surface for the bare dump slope and Fig. 5.3b shows the most critical slip surface for the vegetated dump slope. This increase in the slip surface radius is due to the higher mobilization of the shear strength. The higher FoS is attributed to the enhanced value of cohesion due to the root reinforcement. However, a probabilistic analysis is required to clearly understand the effect of material heterogeneity.

5.3.2 Probabilistic Analysis

A probabilistic analysis was conducted on the vegetated slope to assess the impact of coefficient of variation (CoV) on the stability in terms of mean FoS, P_f, and β. Figure 5.4 depicts the effect of CoV in the material properties (c, ϕ, and γ) of overburden with and without vegetation on the P_f. The solid lines represent the P_f values for the bare slope while the dotted lines represent the P_f value for the dump with roots. It is evident that the probability of failure for the vegetated dump is lower as compared to the bare dump (Fig. 5.5). For bare slope, $P_{f(b)} = 16.7\%$ was obtained for Case 1 with minimum CoV and $P_{f(b)} = 50.7\%$ was obtained for Case 5 having maximum CoV. For the dump with vegetation, the $P_{f(v)} = 14.0\%$ was recorded for Case 1 while $P_{f(v)} = 46.75\%$ was recorded for Case 5. The deterministic FoS is constant for each case (1.115). The influence of material uncertainty on the mean factor of safety is depicted in Fig. 5.6, revealing that an increase in uncertainty corresponds to a decrease in the mean value of FoS.

The PDF of the FoS was plotted for the OB material in Fig. 5.7, using the continuous random variables for different $CoVs$. It can be observed that the PDF shows a widespread with an increase in CoV values. The minimum FoS range was observed under Case 1 (0.75–1.50), while the maximum range was observed under Case 5 (0.25–2.0). To gain a better understanding of the probability of failure for each case, the data from the PDF, relative frequency, and FoS (histogram) were combined and depicted in Fig. 5.8. An arrow marks the curve, indicating the mean factor of safety (FoS) values with densities (gradient of cumulative distribution function, CDF). To

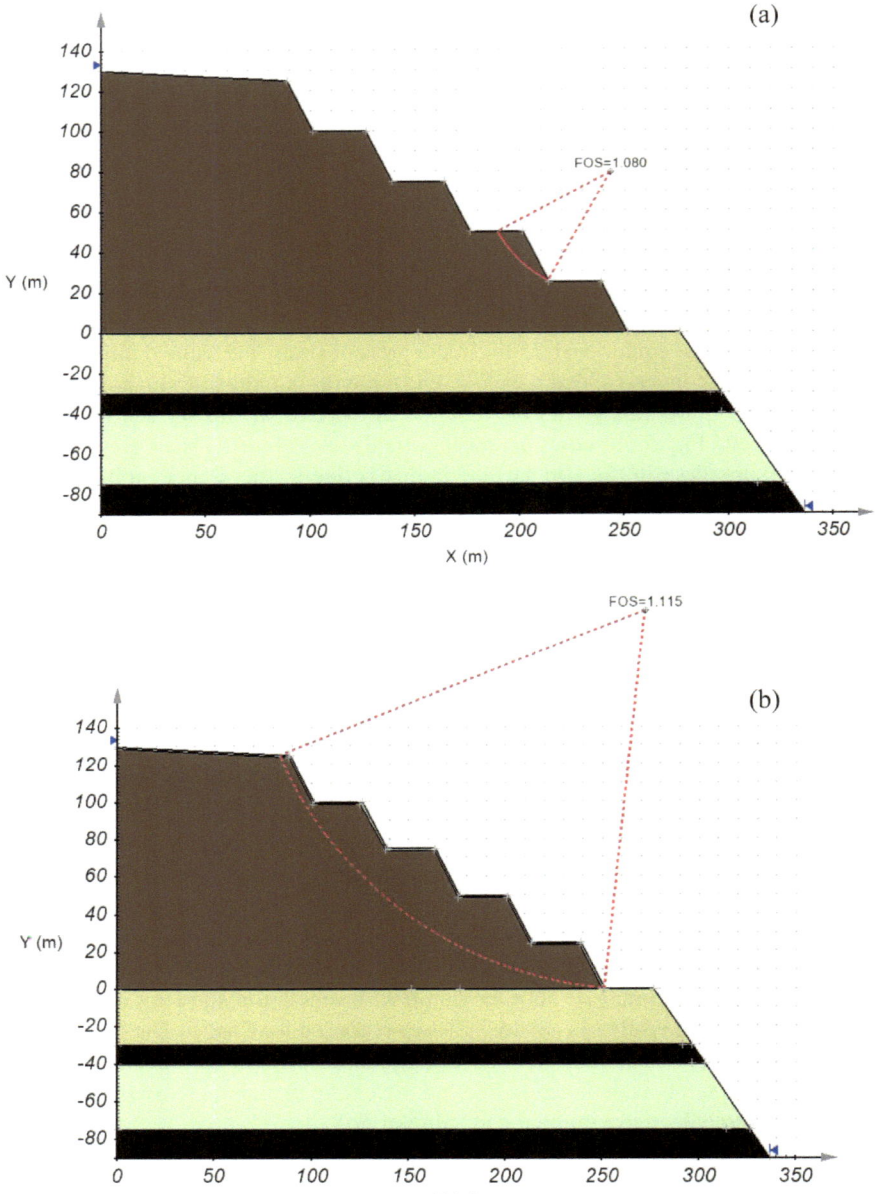

Fig. 5.3 Deterministic slope stability output showing the most critical slip surface in **a** bare slope and **b** vegetated slope

Fig. 5.4 P_f for different combinations of CoVs of the material properties

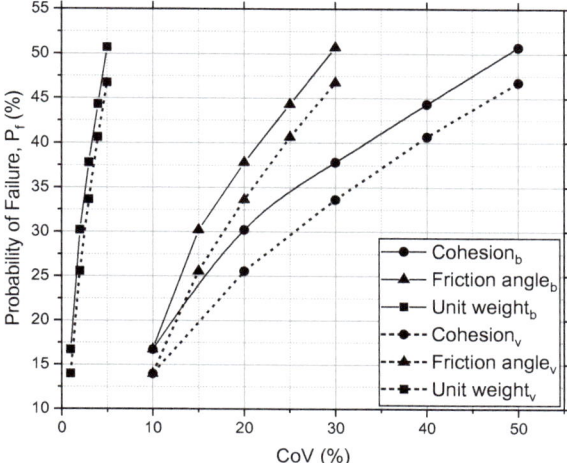

Fig. 5.5 Variation of P_f (%) for different CoV cases for the vegetated slope

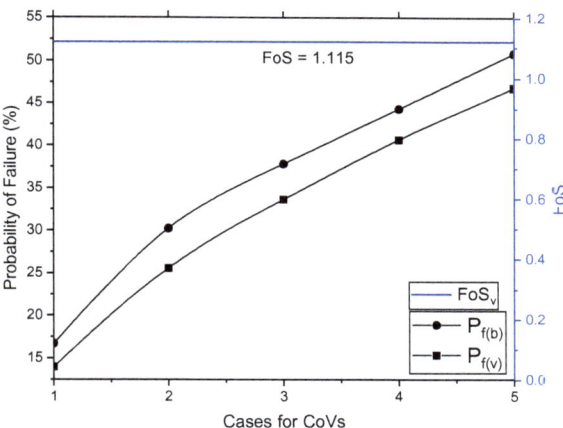

the left of the mean, a straight vertical line shows $FoS = 1$, with the region to the left of the line representing the unstable region where $FoS < 1$. P_f, the probability of failure, is defined as the ratio of the area under the curve for FoS < 1 to the total area under the curve. For Case 1, P_f was calculated to be 14.0%, meaning that 140 out of 1000 similar slopes are expected to fail at some point in their lifetime. This value for the bare slope was 167 in 1000 slopes. In Case 1, the gap between the line and the arrow is the widest (Fig. 5.8a). Increasing the coefficient of variation shifts the line at $FoS = 1$ towards the mean value, reducing the gap and indicating an increased P_f. The gap between the line at $FoS = 1$ and mean FoS is the least in Case 5 with $P_f = 46.75\%$ (Fig. 5.8d).

The cumulative representation can also show the probability of failure for different factors of safety, which was analyzed using the MCM. Figure 5.9 displays the P_f-FoS

Fig. 5.6 Variation of mean
FoS with the material CoV

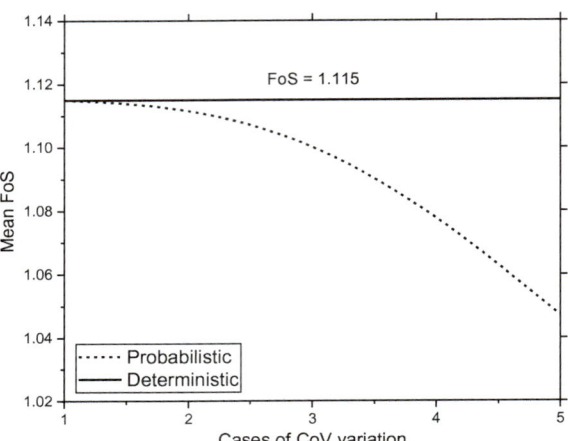

F g. 5.7 *FoS* distribution of
the vegetated coal OB dump
for the different *CoV* cases

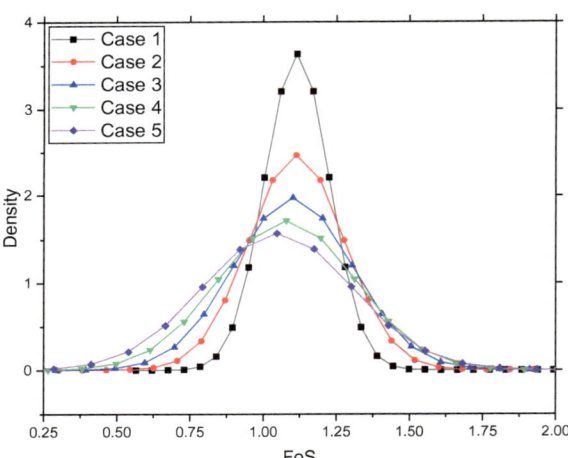

plot for various scenarios, which represent the heterogeneity in the OB material. The
'S' curve obtained for Case 1 had a narrow range of FoS (0.85–1.4) when compared
to Case 5 (0.40–1.70) for various cumulative P_f values. The five different curves
intersected at $FoS = 1.115$, which was also the *FoS* for the deterministic analysis
of the vegetated slope. At the intersection point, the probability of failure was 63%
which was the same as that of the bare slope.

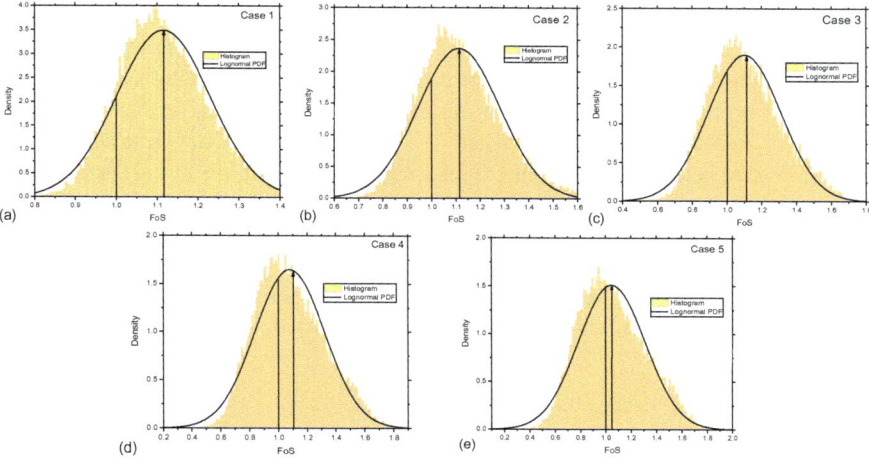

Fig. 5.8 Overlay of relative frequency versus *FoS* and PDF versus *FoS* for different cases: **a** Case 1, **b** Case 2, **c** Case 3, **d** Case 3, and **e** Case 5

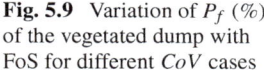

Fig. 5.9 Variation of P_f (%) of the vegetated dump with FoS for different *CoV* cases

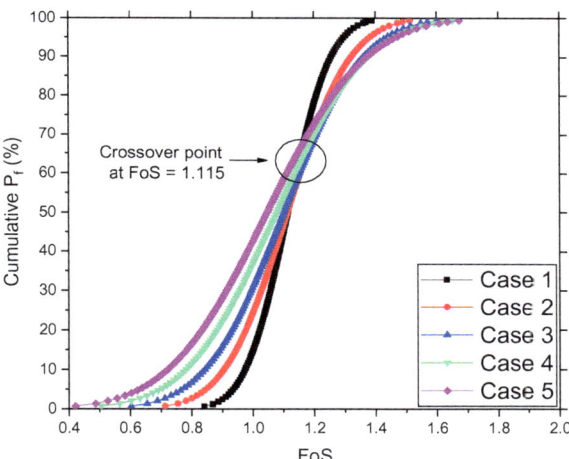

5.4 Conclusions

The present study is focused on studying the effect of enhanced soil cohesion due to the presence of roots on the slope stability of the heterogeneous OB dump. The deterministic analysis performed using the Limit Equilibrium Method yielded a *FoS* = 1.115 which was slightly higher than that obtained on the bare slope (1.08). According to the probabilistic analysis, as the variability in the material increased, the likelihood of failure increases. The reliability index and the material coefficient of variation values had an inverse relationship. The average factor of safety value decreased with increasing uncertainty, and the range of factor of safety values also increased. The relationship between the probability of failure and the factor of safety for the different coefficient of variation cases demonstrated a crossover point at a factor of safety value of 1.115, where the probability of failure was 63.38%. To account for the random dumping of OB on dumps, it is crucial to account for variations in material properties. This necessitates the use of probabilistic analysis to factor in heterogeneity and assess the stability of the slope. This approach may also be relevant in situations where dump characteristics change gradually over time. The limitation of the study lies in the assumption that the material is considered to be homogeneous and that there is no significant alteration in soil density due to root penetration.

References

1. Sun, Y. Z., Fan, J. S., Qin, P., & Niu, H. Y. (2009). Pollution extents of organic substances from a coal gangue dump of Jiulong Coal Mine, China. *Environmental Geochemistry and Health, 31*(1), 81–89.
2. Dary, M., Chamber-Pérez, M. A., Palomares, A. J., & Pajuelo, E. (2010). "In situ" phytostabilisation of heavy metal polluted soils using Lupinus luteus inoculated with metal resistant plant-growth promoting rhizobacteria. *Journal of Hazardous Materials, 177*(1–3), 323–330.
3. Koner, R., & Chakravarty, D. (2016). Characterisation of overburden dump materials: A case study from the Wardha valley coal field. *Bulletin of Engineering Geology and the Environment, 75*, 1311–1323.
4. McQuillan, A., Canbulat, I., Payne, D., & Oh, J. (2018). New risk assessment methodology for coal mine excavated slopes. *International Journal of Mining Science and Technology, 28*(4), 583–592.
5. Tripathi, N., Singh, R. S., & Chaulya, S. K. (2012). Dump stability and soil fertility of a coal mine spoil in Indian dry tropical environment: A long-term study. *Environmental Management, 50*(4), 695–706.
6. Poulsen, B., Khanal, M., Rao, A. M., Adhikary, D., & Balusu, R. (2014). Mine overburden dump failure: A case study. *Geotechnical and Geological Engineering, 32*(2), 297–309.
7. Sahu, K., & Mishra, S. (2018). Scientific study and advice on mitigation of silt generation and boulders & debris flow in the coal mine overburden. *International Journal of Science and Research, 7*, 898–903.
8. Directorate General of Mines Safety (DGMS). (2017). Coal mines regulation (CMR). In Gaz. India Extraordinary. https://www.dgms.net/CoalMinesRegulation2017.pdf.
9. Kumar, A., Das, S. K., Nainegali, L., Raviteja, K., & Reddy, K. R. (2022). Probabilistic slope stability analysis of coalmine waste rock dump. (Under review).
10. Varela, C., Vazquez, C., Gonzalez-Sangregorio, M. V., Leiros, M. C., & Gil-Sotres, F. (1993). Chemical and physical properties of opencast lignite minesoils. *Soil Science, 156*(3), 193–204.
11. Ulusay, R., Arikan, F., Yoleri, M. F., & Çağlan, D. (1995). Engineering geological characterization of coal mine waste material and an evaluation in the context of back-analysis of spoil pile instabilities in a strip mine, SW Turkey. *Engineering Geology, 40*(1–2), 77–101.
12. Kasmer, O., Ulusay, R., & Gokceoglu, C. (2006). Spoil pile instabilities with reference to a strip coal mine in Turkey: Mechanisms and assessment of deformations. *Environmental Geology, 49*(4), 570–585.
13. Keskin, T., Makineci, E. (2009). Some soil properties on coal mine spoils reclaimed with black locust (Robinia pceudoacacia L.) and umbrella pine (Pinus pinea L.) in Agacli-Istanbul. *Environmental Monitoring and Assessment, 159*(1), 407–414.

14. Steiakakis, E., Kavouridis, K., & Monopolis, D. (2009). Large scale failure of the external waste dump at the "South Field" lignite mine, Northern Greece. *Engineering Geology, 104*(3–4), 269–279.
15. Świtoniak, M., Hulisz, P., Różański, S., Kałucka, I. (2013). Soils of the external dumping ground of the Bełchatów open-cast lignite mine. In *Technogenic soils of Poland* (pp. 255–274). Polish Society of Soil Science.
16. Behera, P. K., Sarkar, K., Singh, A. K., Verma, A. K., & Singh, T. N. (2017). Dump Slope stability analysis–a case study. *Journal of the Geological Society of India, 89*(2), 226–226.
17. Masoudian, M. S., Zevgolis, I. E., Deliveris, A. V., Marshall, A. M., Heron, C. M., & Koukouzas. N. C. (2019). Stability and characterisation of spoil heaps in European surface lignite mines: A state-of-the-art review in light of new data. *Environment and Earth Science, 78*(16), 1–18.
18. Bischetti, G. B., Chiaradia, E. A., Epis, T., & Morlotti, E. (2009). Root cohesion of forest species in the Italian Alps. *Plant and Soil, 324*(1), 71–89.
19. Comino, E., Marengo, P., & Rolli, V. (2010). Root reinforcement effect of different grass species: A comparison between experimental and models results. *Soil Tillage Research, 110*(1), 60–68.
20. Rai, R., & Shrivastva, B. K. (2011). Biological stabilization of mine dumps: Shear strength and numerical simulation approach with special reference to Sisam tree. *Environment and Earth Science, 63*(1), 177–188.
21. Babu, G. L. S., Reddy, K. R., & Srivastava, A. (2014). Influence of spatially variable geotechnical properties of MSW on stability of landfill slopes. *Journal of Hazardous, Toxic, Radioactive Waste, 18*, 27–37.
22. Raghuram, A. S. S., & Basha, B. M. (2020). Analysis of rainfall-induced slope failure using Monte Carlo simulations: A case study. In *Geohazards: Proceedings of IGC 2018 Singapore* (pp. 111–127). Springer.
23. Reddy, K. R., Kumar, G., Giri, R. K., & Basha, B. M. (2018). Reliability assessment of bioreactor landfills using Monte Carlo simulation and coupled hydro-bio-mechanical model. *Waste Management, 72,* 329–338.

Chapter 6
Suitability of Bauxite Residue as a Landfill Liner Material—An Overview

Narala Gangadhara Reddy⍟, **Tayyaba Siddiqua, Manikanta Devarangadi, and Chandra Bogireddy**

6.1 Introduction

As development progresses, the need for sustainable materials and better management of solid wastes also increases. Industries, which play a major role in the development of a nation produce tons of waste/byproducts and its disposal is a major concern. The byproducts disposed of pollute land, water and the surrounding environment. To overcome such problems utilization of waste materials according to their properties is a major solution. Aluminium ranks third among the most abundant elements on the Earth and its extraction has highly increased over a period of time. Among the largely produced byproducts, bauxite residue (BR) is a byproduct formed after the production of alumina by Bayer's or sintered process. In Bayer's process bauxite is digested in NaOH at high temperature and high pressure of 250 °C and 52 kg/cm^2 respectively.

It is estimated that BR produced across the globe is more than 120 million tonnes annually. BR is highly alkaline (pH > 10) due to the addition of caustic soda (NaOH) to bauxite ore during the extraction of alumina [1]. The major chemical constituents of BR are Fe_2O_3, SiO_2, Al_2O_3, Na_2O, and CaO. The chemical composition of BR

N. G. Reddy (✉) · T. Siddiqua
Kakatiya Institute of Technology and Science, Warangal 506015, India
e-mail: gn11@iitbbs.ac.in

N. G. Reddy
School of Building and Civil Engineering, Fiji National University, Samabula, Suva 3722, Fiji

M. Devarangadi
Ballari Institute of Technology and Management, Ballari 583104, India

C. Bogireddy
Shantou University, Shantou 515063, Guangdong, China

© The Author(s), under exclusive license to Springer Nature Singapore Pte Ltd. 2024   89
S. K. Das et al. (eds.), *Geoenvironmental and Geotechnical Issues of Coal Mine Overburden and Mine Tailings*, Springer Transactions in Civil and Environmental Engineering, https://doi.org/10.1007/978-981-99-6294-5_6

produced mainly depends on the origin of bauxite, mineral composition, and method of processing [1–3].

Due to the high volume of BR generated its safe disposal becomes a major issue. Generally, BR is discarded in landfills, rivers, and ponds. The construction of tailing dams for the disposal of BR utilises huge acres of land and conventional materials [2]. Moreover, the failure of such dams causes contamination of surface water and land. The average size of BR is around 10 μm causes air pollution and groundwater contamination through leaching [3]. Failure of BR dike in Ajka, Hungary in 2010 discharged about 0.6 million cubic meters of BR, killing 10 people and injuring 120 people [4]. Similar incidents were also reported in India and China [2, 5]. The stockpiling of BR is a major concern worldwide and its utilization is still limited. Thus, to avoid such incidents and preserve natural resources, utilization of BR in a sustainable way becomes inevitable. Considering the need for sustainable utilization of BR, an examination of the properties of BR has been increased. Studies to determine the suitability of BR for various applications increased over a period of time. BR has various applications in construction, contaminated water and soil treatment, recovery of rare earth metals, and catalyst in steel production. The main objective of this chapter is to assess the suitability of BR as compacted landfill liner material through literature. Hydraulic conductivity, the most significant property affecting the performance of liner and other properties like grain size distribution, compaction characteristics, and shear strength were discussed. Further, the potential of BR as a liner with different additives is also presented.

6.2   Landfill Liner

A landfill is an engineered structure consisting of a liner system, waste layer and cover. The cover is on the top which resists the water reaching into the waste layer and the liner at the bottom resists the migration of leachate from reaching the groundwater [6]. Landfills are constructed for the disposal of industrial, construction, demolition, hazardous, municipal wastes and other wastes like coal combustion residues. The liner used in landfill acts as a barrier between waste materials disposed and the surrounding environment to prevent groundwater, land contamination by contaminants, and leachate produced in landfill [7].

Figure 6.1 shows the different types of liner systems. Conventional liners also include geosynthetic clay liners and composite clay liners. Considerable studies were conducted on pond ash, compacted silt loess, cement kiln dust, bentonite embedded zeolite, bentonite with coated gravel, and clay with lime were identified as suitable liner materials as these materials possess low hydraulic conductivity and have good adsorption properties [8]. The liners are then designed with different adaptions like using sand-bentonite mixtures, High-density polyethelyne membrane (HDPE) for waste containment. In recent years due to the scarcity of natural clays, the conventional clay liners are replaced by locally available sand and lateritic soils blended with bentonite and residual soils [9]. Further, the researchers studied the properties of industrial byproducts for their utilization as liners. Industrial byproducts such as steel slag, ground granulated blast furnace slag (GGBS byproduct of iron-steel industries), fly ash and bottom ash (byproduct of thermal power plants) are produced in large quantities worldwide and were studied for their suitability as landfill liners [8–10]. Among these industrial byproducts, BR was also studied for its suitability as a landfill liner. The liners are mainly classified into two types which are further classified based on the materials used to construct them. A single liner is made up of a liner with a leachate collection system above it and the double liner system, as the name suggests is made up of two liners with a leachate collection system at the top and a leakage detection system at the bottom called as drainage layer. Composite liners are those in which two materials of low hydraulic conductivity are used to build a liner system [10]

As the landfill liner is to prevent the migration of leachate from waste into surrounding land and groundwater thus, a material satisfying the standard requirements can be utilized as a liner. The standard requirements for landfill liners are listed in Table 6.1.

The conventional liner's construction becomes costly if the material needed to be imported and transported to the construction site and clay liners are highly susceptible to chemical attacks. However, BR is available in bulk quantities at alumina refineries, highly resistant to chemical attacks and its sorption capacity is very high for many contaminants. The sorption capacity is beneficial to deal with hazardous waste materials [11–13]. Thus, BR has the potential to be used as a landfill liner.

Fig. 6.1 Types of landfill liner systems based on the requirement

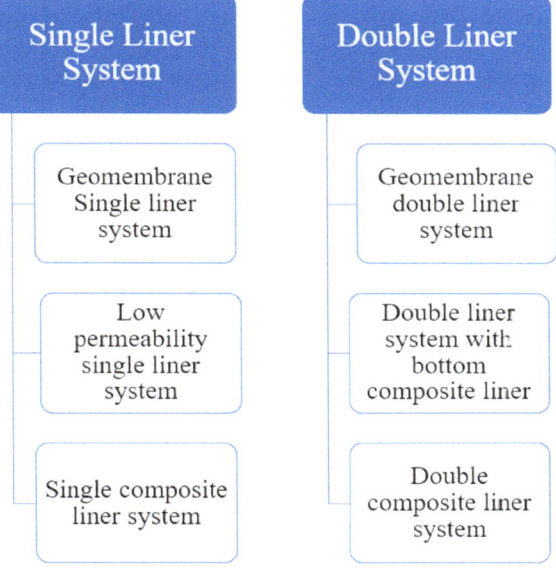

Table 6.1 Standard requirements of a liner material [6, 10]

S. no	Parameters	Standard requirements
1	Hydraulic conductivity	$\leq 1 \times 10^{-7}$ cm/s
2	Unconfined compressive strength	≥ 200 kPa
3	Maximum dry density	≥ 1.7 g/cc
4	Remoulded undrained shear strength	≥ 50 kPa
5	Activity	≥ 0.3
6	Specific gravity	≥ 2.5
7	Volumetric shrinkage strain	$<4\%$
8	Classification	CL and CH CL and CI
9	Atterberg's limits	LL $\geq 20\%$, PI $\leq 7\%$
10	Grain size analysis	Fines content $\geq 20\%$ % gravel ≤ 30 Largest grain size ≤ 63 mm Silt size content $\geq 15\%$ Clay size content $> 20\%$

6.3 Suitability of Bauxite Residue as Landfill Liner

Geotechnical properties of BR around the world are collected and reported in Table 6.2. It is observed that liquid limit and plastic limit are in the range of 21–54% and 16–40% respectively. Most of the BRs fall under the category of silt of low plasticity (ML) as per USCS classification with a high percentage of clay and silt particles. The compaction characteristics (maximum dry density, γ_{dmax}, and optimum water content w_{opt}), and hydraulic conductivity (K) were also presented in Table 6.2.

6.3.1 Compaction Characteristics and Hydraulic Conductivity (K) of Bauxite Residue

The compaction properties of a material depend on the type of compaction, grain size distribution and water. Low permeability can be obtained at higher compaction effort due to the reduction of porosity. Moreover, the lower hydraulic conductivity, lowest values of porosity, and void ratio can be attained when geomaterials are compacted at maximum dry density (γ_{dmax}) at optimum water content (w_{opt}). Devarangadi and Shankar [8] noted that low K values can be obtained at water contents from 0.96% to 4.8% of wet of optimum. The particles defloculate at higher moulding water content and this helps to reduce void space in turn hydraulic conductivity decreases. As the compaction effort moves from dry to wet of optimum moisture content, hydraulic conductivity decreases.

Table 6.2 Geotechnical engineering properties of bauxite residue samples

Source	USCS	Clay (%)	Silt (%)	Sand (%)	LL (%)	PL (%)	PI (%)	γ_{dmax} (g/cc)	w_{opt} (%)	K (cm/s)
ALCOA, San Cibrao bauxite refinery, Spain [13]	ML	40	50		39 ± 2	31 ± 2	8	1.69	7–35	9×10^{-8}
MYTILINEOS alumina refinery, Greece [3]	ML	52	35	13	35	30	5	1.436	29	1×10^{-7}
Guangxi Xingfa Aluminium Electricity Co. Ltd, Baise City, China [14]	–	–	–	–	31.20	–	–	1.44–1.58	25.1–33.5	–
NALCO at Damanjodi Koraput, Odisha, India [15]	ML	22–35	43–76	5–15	21–45	16–36	5–7	1.25–1.7	22–34	5.83×10^{-4} to 0.13×10^{-8}
HINDALCO at Belgaum, Karnataka, India [5]	ML	–	–	–	39	28	11	1.59	34.39	–
ALCAN, Jamaica [16]	ML	20–30	50	20–30	45	36	9	–	–	1×10^{-5} to 1×10^{-7}
Guinea bauxite residue [17]	ML-CL	20	44	20–30	44–66	33–36	11–26	1.55	30	1.4 to 6.7×10^{-6}
UK bauxite residue [2]	MH	20	80	0	54	40	14	1.75	–	2 to 10×10^{-7}
Hindalco Plant, Renukoot, India [15]	ML	29–39	43–57	10–14	40–45	30–35	5–14	1.45–1.64	33	2 to 3×10^{-6}
Hindalco Plant, Muri, India [18]	ML	32	51	17	39.89	36.08	3.81	1.52–1.59	–	5.3×10^{-4} to 8.7×10^{-5}

Rubinos et al. [13] established the dry density water content relation of a BR sample and observed that the compaction curve followed the same trend and shape as that of clayey soils. The same study also demonstrated that the least K value (<1.8 × 10^{-7}) was obtained for the water content between 28.5 and 32% (0.4–4% wet of optimum) and the γ_{dmax} obtained was 1.69 g/cc. A study by Panda et al. [19, 20] showed that BR sample had a γ_{dmax} of 1.87 g/cc at 22.6% w_{opt}. However, available literature shows that most of the BR samples failed to meet the standard requirement of γ_{dmax} (>1.7g/cc), this is due to smaller particle size, surface charge, high amount of iron oxide (55%) and highly alkaline least to low dry density and high porosity. So, to overcome this Reddy et al. [20, 21] suggested that BR should be compacted at higher compaction energy or admixing suitable additive to BR can be adopted to enhance the compaction properties of BR.

The studies were conducted on BR samples from China and India, by adding additives to BR, and it is noted that γ_{dmax} values increased with an increase in the amount of additives. For the China BR sample, the values of γ_{dmax} were in the range of 1.44–1.58 g/cc. The additives used were lime and cement. For the BR of Karnataka, India, it has been observed that the γ_{dmax} increased with increase in gypsum content. The highest value of γ_{dmax} obtained was 1.66 g/cc for sample with 80% BR, 20% FA and 1% of gypsum with w_{opt} of 31.52%.

Zevgolis et al. [3] have studied the role of fly ash, bentonite, and cement on permeability of BR. The tests were conducted on 3 sets of mixtures per additive consisting of 5, 10, and 15% of additive by weight. The results demonstrate that bentonite and cement admixed samples had a positive effect on permeability to meet the standard requirement. It was found that fly ash is not a suitable additive to decrease the permeability of BR sample. As in all the three samples admixed with fly ash, there was a decrease in γ_{dmax} values, whereas the values of w_{opt} are higher than that of raw BR. Regarding the permeability, fly ash admixed BR have not met the specifications. To the samples admixed with bentonite there was a rise in γ_{dmax} and fall in w_{opt} with an increase in bentonite content. Concerning permeability, the values of K obtained met the standard required criteria. For the cement admixed samples there was a significant increase in γ_{dmax} with an increase in the cement content and the values of K were fulfilling the standard requirement at different vertical pressures i.e., $\leq 1 \times 10^{-7}$ cm/s. For the Indian BR sample the study by Panda et al. [19] demonstrate that the values of permeability of bioneutralized BR have decreased when compared to raw BR sample. The values were significantly reduced from 5.13 × 10^{-8} to 1.23 × 10^{-8} cm/s.

6.3.2 Index Properties of Bauxite Residue

Index properties are an important factor to deal with the performance of the landfill liner material. The index properties rely on the size of particles. As the fines' content increases the liquid limit also increases as a result the hydraulic conductivity significantly decreases [8]. This is due to small pores and electrical charge of the fine particles. The material to be used as a liner should contain fine content > 20% in order to meet the standard requirement of hydraulic conductivity [10]. The sand content of the samples is in the range of 0–30%, clay content from 22 to 52% and silt content from 35 to 76%. Sand offers resistance to volumetric shrinkage and provides sufficient strength [10]. It is observed that the clay content in all the samples is >20% satisfying the standard requirement. As per studies, the BR sample contains silt and clay of higher percentages which together contribute 88% of the entire BR particles [22]. As such high fine contents in all the BRs irrespective of origin are favourable to achieve lower K values and thus reduction of leachate migration [10]. Although, soils with high swelling potential show apparently a low-hydraulic conductivity these soils crack upon drying due to shrinkage and lead to permeation of the leachate and contaminate the environment. In this regard, the use of BR as a liner material gains merit as it is known as a non-swelling material. The values of liquid limit, plastic limit and plasticity index are summarized in Table 6.2.

6.3.3 Strength Characteristics of Bauxite Residue

A landfill liner must possess sufficient strength to resist sliding on the slope. The compressive strength depends upon compaction characteristics, and higher densities of material leads to higher compressive strength as the material gets closely packed. Rubinos et al. [13] performed a direct shear test on BR sample and reported friction angle and cohesion as 38° and 79.5 kPa respectively, which are much higher when compared to compacted clay. The higher values of frictional angle and cohesion have a positive effect on strength. Unconfined compression tests (UCS) conducted on Indian BRs were in the range of 136.5–240 kPa [2].

Ou et al. [14] conducted experiments to improve the strength of BR by admixing cementitious materials. The tests were carried out on samples with different ratios of BR, tailings mud, (mud produced by grinding and washing of bauxite ore) quicklime (CaO) and cement with different curing periods. UCS values of these samples were in the range of 2.68–4.54 MPa which is more than ten times of the required strength. The highest UCS was obtained for the sample 1:0.2 (w/w waste to cementitious material ratio) with 28 days of curing. Though the studies were conducted to assess the suitability of BR as a subgrade material, it can be concluded that the UCS of BR can be enhanced with cementitious additives and curing time is another important aspect. The studies conducted by Chandra and Krishnaiah [5] by replacing BR with 10, 20, and 30% fly ash (FA) by its dry weight and adding 0.5 to 1% of gypsum with a varying curing period of 1, 7, and 28 days. The results demonstrate that the addition of FA to BR enhanced UCS to a greater extent and higher UCS values were obtained for 20% of FA. In both cases, it has been observed that higher UCS can be obtained if the curing period is longer. Thus, the strength parameters of BR can be improved with an appropriate percentage of additives. From the studies conducted by Panda et al. UCS obtained for bioneutralized BR was 298.6 kPa, which is not only much higher than clayey soils and ≥200 kPa i.e., minimum UCS required for a liner material. Therefore, BR strength can be improved with the appropriate use of additives such as other industrial waste or cementitious materials for better waste management and its use in landfill liner applications by satisfying the design criteria.

6.3.4 Desiccation Cracking of Bauxite Residue

Desiccation is the formation of cracks, which has many adverse effects on the performance of the liner material. This is because cracks could dramatically increase the permeabilty of the geomaterial and destroy the main idea of providing a liner to the landfill. The desiccation cracks may lead to various geoenvironmental problems like contamination of groundwater, retaining less water in drought conditions, and reduction of soil strength [6].

Rubinos et al. [13] studied the effect of desiccation on permeability and particle size distribution. The samples were dried at 3 different conditions i.e., oven dried at 60 and 110 °C and air dried for 21 days. The results demonstrate that when samples are dried there was a decrease in clay content (~7–25 times) and fine silt (~2–5 times) with an increase in coarse sand fractions (~14–20 times). Even though there was a change in particle size distribution there was no change in USCS classification and Atterberg's limits. Whereas, an increase in hydraulic conductivity on desiccation was found. This was observed by testing a compacted BR specimen subjected to two consecutive wet-dry cycles with hydraulic conductivity determined after each cycle. Even though there were no cracks observed, there was an increase in hydraulic conductivity due to wet-dry cycling. But the increase in hydraulic conductivity upon desiccation was less than the increase found in clay liners. This is due to the presence of minerals in BR which possess low swelling potential and crystallinity.

6.3.5 Adsorption and Leaching Characteristics of Bauxite Residue

Besides favourable geotechnical properties, BR has good adsorption properties for the removal of heavy metals, dyes, phosphate, nitrate, fluoride, and arsenic. Studies of Coruh et al. (2010) [23] showed that BR is an effective adsorbent for the removal of zinc ions from zinc leach residue. As the BR contains metallic oxides and hydroxides of Al and Fe which help in metal adsorption. Further BR has higher iron content which resists chemical attacks [11]. As per the studies conducted by Rubinos et al. [12, 13] it was found that BR is an effective adsorbent of toxic metals like Hg and As and also effective to retain Hg and As from aqueous system. Moreover, the adsorption and retention of heavy metals by BR can be effectively applied in various engineering applications like for decontamination of metal rich industrial effluents or acid mine drainage, as a chemical barrier material to mitigate toxic elements, as an integrant of liner systems for Hg-rich waste disposal units.

Any waste may contain impurities like heavy metals and metalloids, leaching characteristics of such waste play a vital role to assess its environmental compatibility, possible threat to humans and living organisms and its potential to diminish the quality of water. Available literature shows that BR contains elements like As, Zn, Mn, Cd, Pb, Hg, Na, Ca, Cu, Si, Ar. TCLP (Toxicity Characteristics Leaching Procedure) was conducted by various researchers to determine the heavy metals present in BR leachate. TCLP tests conducted by Rubinos et al. [23] on BR liners using water, acetic acid and 5% $CaCl_2$ demonstrated that concentrations of Al and Cr in BR leachate exceeded the WHO drinking water standards. Nonetheless, these metal outputs decreased upon the removal of suspended solids from BR liners. From the leaching studies conducted by Chandra and Krishnaiah [5] for the best combination of 80% BR + 20% FA + 1% gypsum and Zhao et al. [24] for raw BR it was found that the concentration of all the heavy metals in BR leachate were within TCLP regulatory limits. Another study conducted by Li et al. [25] demonstrated that the contents of toxic elements leached were within regulatory standards. Although in the above studies the toxic elements leached were within standards it has to be noted that the method of extraction adopted by refinery, chemical composition of BR directly affects the concentrations of the elements in leachate [2]. Moreover, the leaching of toxic elements and sorption/adsorption properties of BR are pH dependent. Thus, neutralizing BR with strong acids not only reduces the availability of toxic metals and metalloids but also enhances its sorption capacities [26, 27]. The addition of gypsum also reduces the toxicity of BR leachate through sequestration of atmospheric carbon [27]. Mishra et al. [28] studied leaching characteristics of ameliorated BR with gypsum, OPC and GGBS and found that both untreated BR treated BR is nontoxic as toxic element contents were within USEPA prescribed limits. Qu et al. [29] found that leaching of toxic metals from BR can be potentially reduced by bioleaching i.e., leaching of metals with heterotrophic fungus. TCLP tests conducted by Chandra et al. [30] on fly ash-gypsum geopolymer composites demonstrated that for BR and 80% BR-20% fly ash, the toxic elements were within permissible limits. Thus, toxicity of BR can be treated to make it potential for safe usage. However, site conditions may not be similar to that of lab environment which makes it challenging to predict the leaching behaviour of BR.

6.4 Future Research Needs

The suitability of BR as the liner material reviewed in this study is based on the experiments conducted by various researchers in a controlled laboratory environment which may not replicate the field conditions. There might be a possibility of facing some challenges in field scenarios. On contrary, BR in the fields may undergo different chemical and environmental conditions which might not get imbibed in the laboratory tests. Though there are efforts made by many researchers to evaluate the properties of BR, still there is need to evaluate the BR completely to make it more adaptable. From the review of literature, the following research gaps are found to work further to explore new avenues for better utilization of BR as a liner material:

- Effect of additives on hydraulic conductivity under different environmental conditions need to be evaluated so that BR can be utilized as an effective liner under different climatic and alkaline/acidic conditions.
- Studies on desiccation cracking and water retention properties of BR are scarce. As desiccation affects many engineering properties (such as strength and stability), there is an urgent need to study to enhance its utilization with and without additives.
- Long term durability and short term stability of BR, when admixed with additives, are rarely explored.
- There is a need to explore the volumetric shrinkage strain and diffusion coefficient of BR.
- Novel and innovative engineering applications should be explored to make use of BR in the most sustainable way and to reduce reliance on natural materials by fulfilling the standard requirements without compromising serviceability.

6.5 Concluding Remarks

A brief review of the geotechnical and geoenvironmental engineering properties of BR for landfill liners suitability is presented. From this review the following conclusions can be drawn:

- The particle size distribution of BR shows that silt and clay are about 80%. The clay content in the samples is much higher than ($\geq 10\%$) the recommended percentage for clay liner application.
- The liquid limit and plasticity index of BR samples were in the range of 21–45% and 5–7% respectively. The liquid limit value of BR meets the standard requirement i.e., $\geq 20\%$. The plasticity index of most of the samples also meets the recommended value of $\geq 7\%$. It should be noted that the properties of BR largely vary based on the type of the ore chief mineral, type of processing and handling of residues.
- The values of hydraulic conductivity were in the range of 5.83×10^{-4} to 0.13×10^{-8} cm/s. Cementitious additives can be used to reduce hydraulic conductivity to satisfy the requirement of acceptable value (i.e., $\leq 1 \times 10^{-7}$ cm/s).
- The maximum dry density of BR samples are observed in the range of 1.25–1.7 g/cc with or without various additives and this is mainly due to the smaller particle sizes of BR. Though maximum dry density values do not meet the value ≥ 1.7 g/cc as per standards the permeability values obtained at this density were fulfilling the standard requirements for landfill liners.

In general, the BR can be utilized as a landfill liner as the majority of properties meet the standard requirements. Though some minor properties of BR from different origins may not fulfil the requirements can be improved to meet the standards with additives/amendments or bio-neutralization to satisfy requirements. Besides, the strength requirements, environmental safety, and health protection are crucial, so for an effective liner, BRs must undergo a careful evaluation according to the landfill site conditions like landfill waste type, hydrogeological, climatic conditions, and contaminant concentrations.

References

1. Gräfe, M., Power, G., & Klauber, C. (2011). Bauxite residue issues: III. Alkalinity and associated chemistry. *Hydrometallurgy, 108*(1–2), 60–79.
2. Reddy, P. S., Reddy, N. G., Serjun, V. Z., Mohanty, B., Das, S. K., Reddy, K. R., & Rao, B. H. (2021). Properties and assessment of applications of red mud (bauxite residue): Current status and research needs. *Waste and Biomass Valorization, 12*(3), 1185–1217.
3. Zevgolis, I. E., Tsiavos, C., Papangelis, K., Gaidajis, G., & Xenidis, A. (2021). Utilization of bauxite residue as a liner: Permeability improvement using additives. *International Journal of Geosynthetics and Ground Engineering, 7*(1), 1–8.
4. Mayes, W. M., Jarvis, A. P., Burke, I. T., Walton, M., Feigl, V., Klebercz, O., & Gruiz, K. (2011). Dispersal and attenuation of trace contaminants downstream of the Ajka bauxite residue (red mud) depository failure, Hungary. *Environmental Science and Technology, 45*(12), 5147–5155.
5. Chandra, K. S., & Krishnaiah, S. (2022). Strength and leaching characteristics of red mud (bauxite residue) as a geomaterial in synergy with fly ash and gypsum. *Transportation Research Interdisciplinary Perspectives, 13*, 100566.
6. Mei, G., Kumar, H., Reddy, N. G., Huang, S., Balaji, C. R., Sadasiv, S. G., & Zhu, H. H. (2020). Evaluating suitability of geomaterials-amended soil for landfill liner: A comparative study. *Journal of Hazardous, Toxic, and Radioactive Waste, 24*(4), 04020052.
7. Kalkan, E. (2006). Utilization of red mud as a stabilization material for the preparation of clay liners. *Engineering Geology, 87*(3–4), 220–229.
8. Devarangadi, M., & Shankar, U. M. (2020). Correlation studies on geotechnical properties of various industrial byproducts generated from thermal power plants, iron and steel industries as liners in a landfill-a detailed review. *Journal of Cleaner Production, 261*, 121207.
9. Devarangadi, M., & Masilamani, U. S. (2020). Use of sawdust blended with bentonite and cement mixtures to retain diesel oil contaminants as a liner in a landfill. *Indian Geotechnical Journal, 50*(3), 485–504.
10. Emmanuel, E., Anggraini, V., & Gidigasu, S. S. R. (2019). A critical reappraisal of residual soils as compacted soil liners. *SN Applied Sciences, 1*(5), 1–24.
11. Hua, Y., Heal, K. V., & Friesl-Hanl, W. (2017). The use of red mud as an immobiliser for metal/metalloid-contaminated soil: A review. *Journal of Hazardous Materials, 325*, 17–30.
12. Rubinos, D. A., & Spagnoli, G. (2019). Assessment of red mud as sorptive landfill liner for the retention of arsenic (V). *Journal of Environmental Management, 232*, 271–285.
13. Rubinos, D., Spagnoli, G., & Barral, M. T. (2015). Assessment of bauxite refining residue (red mud) as a liner for waste disposal facilities. *International Journal of Mining, Reclamation and Environment, 29*(6), 433–452.
14. Ou, X., Chen, S., Jiang, J., Qin, J., & Zhang, L. (2022). Reuse of red mud and bauxite tailings mud as subgrade materials from the perspective of mechanical properties. *Materials, 15*(3), 1123.
15. Reddy, N. G., Singh, N. R., & Rao, B. H. (2020). Application of biopolymers for improving strength characteristics of red mud waste. *Environmental Geotechnology* https://doi.org/10.1680/jenge.19.00018.

16. Wagh, A. S. (1987). Settling properties of dilute Bayer process muds of alumina industry in Jamaica. *Particulate and Multiphase Processes, 3,* 461–469.

17. Gore, M. S., Gilbert, R. B., McMillan, I., & Parks, S. L. I. (2016). Geotechnical characterization of compacted bauxite residue for use in levees. In *Geo-Chicago* (pp. 299–310).

18. Alam, S., Das, S. K., & Rao, B. H. (2017). Characterization of coarse fraction of red mud as civil engineering construction material. *Journal of Cleaner Production, 168,* 679–691.

19. Panda, I., Jain, S., Das, S. K., & Jayabalan, R. (2017). Characterization of red mud as a structural fill and embankment material using bioremediation. *International Biodeterioration & Biodegradation, 119,* 368–376.

20. Reddy, N. G., & Rao, B. H. (2018). Compaction and consolidation behaviour of untreated and treated waste of Indian red mud. *Geotechnical Research, 5*(2), 106–121.

21. Reddy, N. G., Rao, B. H., & Reddy, K. R. (2019). Chemical analysis procedures for determining the dispersion behaviour of red mud. In A. Agnihotri, K. Reddy, & A. Bansal (Eds.), *Recycled waste materials. Lecture notes in civil engineering.* (vol. 32, pp. 19–26).

22. Çoruh, S., & Ergun, O. N. (2010). Use of fly ash, phosphogypsum and red mud as a liner material for the disposal of hazardous zinc leach residue waste. *Journal of Hazardous Materials, 173*(1–3), 468–473.

23. Rubinos, D. A., Spagnoli, G., & Barral, M. T. (2016). Chemical and environmental compatibility of red mud liners for hazardous waste containment. *International Journal of Environmental Science and Technology, 13,* 773–792. https://doi.org/10.1007/s13762-015-0917-8

24. Zhao, Y., Wang, J., Liu, C., Luan, Z., Wei, N., & Liang, Z. (2009). Characterization and risk assessment of red mud derived from the sintering alumina process. *Fresenius Environmental Bulletin, 18*(6), 989–993.

25. Li, S., Zhang, Y., Feng, R., Yu, H., Pan, J., & Bian, J. (2021). Environmental safety analysis of red mud-based cemented backfill on groundwater. *International Journal of Environmental Research and Public Health, 18,* 8094.

26 Joseph, C. G., Taufiq-Yap, Y. H., Krishnan, V., & Puma, G. L. (2020). Application of modified red mud in environmentally-benign applications: A review paper. *Environmental Engineering Research, 25*(6), 795–806. https://doi.org/10.4491/eer.2019.374.

27. Mayes, W. M., Burke, I. T., Gomes, H. I., Anton, Á. D., Molná, M., Feigl, V., & Ujaczki, É. (2016). Advances in understanding environmental risks of red mud after the Ajka Spill, Hungary. *Journal of Sustainable Metallurgy, 2,* 332–343. https://doi.org/10.1007/s40831-016-0050-z.

28. Mishra, M. C., Reddy, N. G., & Rao, B. H. (2023). Geoenvironmental characterization of bauxite residue ameliorated with different amendments. *Journal of Hazardous, Toxic, and Radioactive Waste, 27*(2), 04022048.

29. Qu, Y., Lian, B., Mo, B., & Liu, C. (2013). Bioleaching of heavy metals from red mud using Aspergillus Niger. *Hydrometallurgy, 136,* 71–77.

30. Chandra, K. S., Krishnaiah, S., Reddy, N. G., Hossiney, N., & Peng, L. (2021). Strength development of geopolymer composites made from red mud–fly ash as a subgrade material in road construction. *Journal of Hazardous, Toxic, and Radioactive Waste, 25*(1), 04020068.

Chapter 7
Fly Ash Based Geopolymer Modified Bitumen (GMB) binder—An Overview

Bojjam Sravanthi◉ and N. Prabhanjan◉

7.1   Introduction

Viscoelastic and thermoplastic properties of bitumen are important as these passed on to asphalt pavements in which it is utilized. Bitumen is a material that is employed in construction. Asphalt mixtures suffer degradation well in advance of the end of their useful lives as a result of both increased traffic and increased environmental impacts [1]. Furthermore, additives are required to improve the strength features of the bitumen binder, such as resistance to rutting and fatigue, as well as resistance to low temperature cracking. According to this viewpoint, polymer modification of the asphalt binder could be a cost-effective approach for reducing pavement maintenance and rehabilitation.

Modified bituminous binder plays a crucial role in various infrastructure projects, particularly in the construction and maintenance of roads, highways, and airports. It is created by modifying conventional bitumen with additives or polymers to enhance its performance characteristics. Modified bitumen offers enhanced resistance to aging and deterioration caused by factors like traffic loads, temperature variations, and weathering. The additives or polymers in the binder improve its elasticity, flexibility, and resistance to cracking, rutting, and fatigue [2–5]. As a result, the modified binder provides better durable pavement surfaces. Bituminous binders tend to become brittle in cold weather and soft in hot weather, leading to cracking or deformation of the pavement [6–8]. By modifying the binder, it becomes more resistant to extreme weathers. It remains flexible at low temperatures, reducing the risk of cracking, and maintains stability at high temperatures, minimizing deformation [4, 8, 9]. Modified

B. Sravanthi (✉)
Kakatiya Institute of Technology and Science, Warangal 506015, India
e-mail: bsr.ce@kitsw.ac.in

N. Prabhanjan
Department of Civil Engineering, SR Engineering College, Warangal 506371, India

binders have better adhesion properties, enabling them to bond more effectively with aggregate materials [2, 10]. This enhanced adhesion contributes to the formation of a stable pavement structure, reducing the occurrence of delamination and aggregate loss. Moreover, modified binders exhibit improved cohesion, allowing the asphalt mix to withstand traffic loads and prevent rutting or shoving. Water intrusion is a significant cause of pavement deterioration. As per the research investigations in pavement designs, modified bituminous binders help to create a more impermeable asphalt mix, reducing the penetration of water into the pavement layers [4, 11, 12]. This improves the overall durability and performance of the pavement, as it minimizes the damage caused by moisture-related issues i.e., stripping and raveling [2, 7]. Moreover, modified binders have improved resistance to rutting due to their enhanced viscosity and elasticity. They can withstand higher traffic volumes and heavy axle loads, maintaining the pavement's integrity and smoothness over an extended period. Modified binders often exhibit improved workability during the construction process [3, 11]. They are easier to handle and mix, ensuring better coating and adhesion to the aggregate particles. This results in a more uniform and homogeneous asphalt mix, leading to improved pavement quality. Some types of modified binders, such as polymer-modified bitumen, offer benefits in terms of sustainability and recycling. They can promote the use of reclaimed asphalt pavement (RAP), a process where old asphalt is recycled and reused [11]. The modified binders help in rejuvenating the aged RAP material, improving its properties and enabling its reuse in new pavement construction [3, 13]. In general, the use of modified bituminous binders in infrastructure projects is essential for achieving long-lasting, high-performance pavements. The improved durability, temperature resistance, adhesion, moisture resistance, rutting resistance, workability, and sustainability aspects make them a valuable component in modern pavement engineering.

To improve the strength of bitumen binder, it must be modified using additives that increase bitumen binder strength. The current research focuses on bitumen binder (traditional asphalt binders) modification with a geopolymer, which plays a vital role in increasing strength, durability, and reducing deformation [2, 3, 7, 13]. Industrial waste materials such as coal mines, GGBS, red mud, CKD, LKD, and RHA are strengthened by the production of this geopolymer. Furthermore, the use of the geopolymer material promotes the utilization of by-products (industrial waste) in pavement construction, reducing the need for landfills. This geopolymer modified bitumen binder (GMB) blend reduces carbon emissions while increasing chemical and thermal effects. Geopolymer exhibits good mechanical qualities at both regular and severe temperature settings. Not only does it boost the strength of (HMA) the hot mix asphalt pavement, but it also increases its durability (long-term performance). The major goals of this chapter are to describe various research findings on geopolymer materials used in flexible pavements, to develop the geopolymer with precursor materials, and to apply the geopolymer modified bitumen binder in asphalt pavements.

7.2   Geopolymer Modified Binder (GMB)

The geopolymer concept was developed in the 1970s, but rapid research began in the year 2000. Figure 7.1 depicts the progress of research on geopolymer materials through 2022 (data source: various studies, geopolymer materials improve the strength, stability, and longevity of bitumen binder and reduce deformation in flexible pavements).

According to new discoveries in material science, geopolymer mix is the most effective material in terms of strength, durability, and sustainability [16] Joseph Davidovits pioneered the use of the geopolymer in 1991. Geopolymers are inorganic, non-crystalline polymers with alumina-silicate chains that are covalently connected. In the geopolymer mix, sodium hydroxide (NaOH) or potassium hydroxide (KOH) and sodium silicate (Na_2SiO_3) or potassium silicate (K_2SiO_3) act as activators. Examples of industrial waste materials that can be used in the GMB include fly ash (FA), metakaolin (MK), red mud (RM), GGBS, cement kiln dust (CKD), and lime kiln dust (LKD) [17]. The geopolymer gel with the bitumen binder must be added to the GMB in order to prepare it. This geopolymer is considered an environmentally beneficial material due to its lower CO_2 emissions. The main advantage of this geopolymer is that it reduces the use of natural materials while increasing the use of materials derived from industrial waste [15, 18]. Table 7.1 shows the general chemical compounds of various precursor materials, and these chemical compositions may vary depending on the source.

7.2.1   Preparation Process of GMB

Precursor materials (Fly ash, GGBS, MK, RHA, LKD, CKD, etc.) and alkali activated solution are required for the development of the geopolymer. The first process in the production of the geopolymer [19] is the use of a liquid alkali activator (LAA), which contains NaOH + Na_2SiO_3 in a 1:2 ratio. The alkali activator and fly ash are mixed

Fig. 7.1   Year wise increases of publications on geopolymer materials (*Data source* Tang et al. [14])

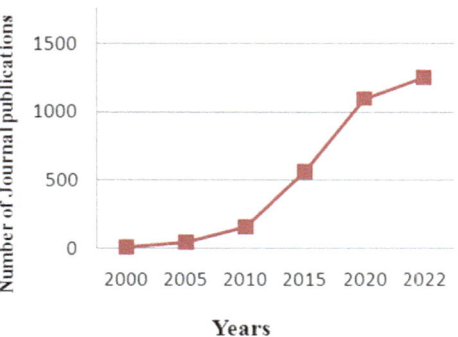

Table 7.1 Chemical composition of different precursor materials (Komnitsa and Zaharaki [10])

Precursor materials	% of Chemical compounds							
	SiO$_2$%	Al$_2$O$_3$%	Fe$_2$O$_3$%	CaO%	Na$_2$O%	K$_2$O%	MgO%	SO$_3$%
Fly ash (FA)	43–61	13–33	3–7	1–22	0–0.95	0–1.54	0.47–4.5	0.6
GGBS	33–42	11–22	0.41–0.71	28–44	0.21–0.4	0.31–1.02	6–13	1.8
Silica fume	82–94	0.25	0.54	0.21	0.22	0.55	0.5	0.16
Rice husk ash (RHA)	88–95	0–0.54	0–0.31	0–0.92	0–0.12	2.3–2.65	0–0.18	–
Cement kiln dust (CKD)	11.70	2.95	2.6	46.5	2.25	4.32	0.67	12.35
Meta kaoline (MK)	38–62	33–54	0.35–2.0	0.06–0.11	–	0.4–1.2	0.07–0.12	–
Red mud (RM)	5–28.35	10.2–35.05	30–60	1.5–15	3–20.2	0–3.5	0–0.32	0–13
Granite waste (GW)	60–75	10–20	1–10	1–2	1–4	3–6	0.3–3.5	0–0.3

well to form the geopolymer gel [20], and the bitumen is heated to 150 °C while the geopolymer gel is mixed with the bitumen binder (VG10 and VG30). Blend the bitumen with the geopolymer gel at 150^{0}C at a mixing speed of 1000 ± 10 rpm [4]. Figure 7.2 depicts a step-by-step approach to prepare the geopolymer modified binder.

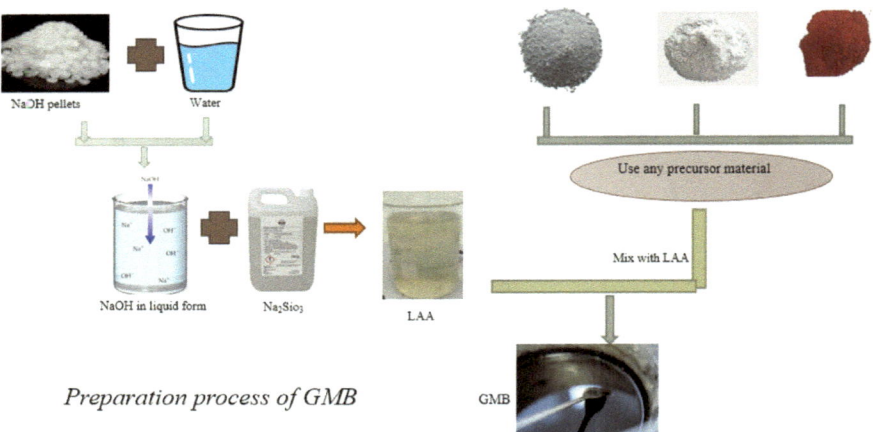

Preparation process of GMB

Fig. 7.2 Preparation of geopolymer modified bitumen binder

7.2.2   Chemical Composition

The geopolymer contains two components, a chemical activator and aluminosilicate. These components develop a chain system in the geopolymer material by mix with the precursor materials. The formation of the geopolymer depends on the alumina and silica ratio of precursor materials, and ratio of NaOH and Na_2SiO_3 [12]. More over the strength properties of the geopolymer are affected by the S/L ratio (solid to liquid ratio), at the range between 0.35 and 1.25 [21]. The mineral content of the precursor material will affect the reaction process and physical properties of GMB (Geopolymer Modified Bitumen). In the fly ash based geopolymer material the ratio of S/L is 0.18 to 0.6, according to this case the 0.18 ratio represents the lower strength compared to the 0.6 ratio [22].

7.3   Influence of Geopolymer Modified Binder (GMB)

7.3.1   Influencing the Physical Properties

According to the research conducted by Ali et al. [9], there is an increase in the softening point value of binder with an increase in the percentage of the modified binder content. Furthermore, the findings of this research indicate that the value increases up to 5% but then begins to decrease after 7% [23, 24]. The specific gravity of bitumen is an important property that helps in assessing and characterizing its quality and performance [24]. Observed by many investigations the specific gravity value of the binder increases with the modified binder content [2, 3, 5, 10, 24]. The importance of viscosity in bitumen lies in its influence on workability, temperature susceptibility, aggregate coating, mixing and pumping operations, and quality control. Understanding and controlling bitumen viscosity help achieve the desired performance characteristics in asphalt pavements, promoting durability, strength, and overall quality [9, 25]. Ali et al. [9] and Ibrahim et al. [24] researched the physical properties of the geopolymer modified binder and discovered that there is a sudden drop in the viscosity of the binder when increasing the temperature from 135° to 165° C, and GMB has a higher viscosity than the virgin binder [21, 23, 24]. According to Al-Mansob et al. [25] the polymer modified binder has the higher viscosity value than the base binder [25]. When it comes to the storage stability of bitumen, several factors can influence its performance and quality over time, such as temperature, contamination, aging and expose to air and moisture. Ibrahim et al. [24] investigated on the storage stability of the geopolymer modified asphalt, and obtained good storage stability values [24, 26].

7.3.2 Influencing the Rheological Properties

Ali et al. [4] used a Dynamic Shear Rheometer (DSR) to investigate the rheological properties of the virgin and geopolymer modified binder, such as the viscous and elastic behavior of the binders—both virgin and modified—at two different temperature conditions, one intermediate and another at the high-in-service temperature [4, 27]. Furthermore, according to the code (AASHTO T315), dynamic shear rheometers evaluate rheological variables such as G* (Complex shear modulus) and (phase angle) of asphalt at the desired frequency and temperatures. Walubita (2011) [4, 28] drew master curves based on G*/sin values; in this study, he used a 25 mm dia plate with a 1 mm gap for intermediate temperatures to conduct frequency sweep tests on bitumen binder, both conventional and polymer modified, at 9 frequencies ranging from 1 to 100 rad/s. Arey et al. studied those rheological properties of SBSMB with different percentages 3, 5, 7 and 9%, and according to him there is an increase in the complex shear modulus value with higher percentage of SBS [27, 29, 30].

   According to Ali et al. (2007), the increase in the complex shear modulus values with geopolymer concentration in the bitumen binder is attributable to an increase in binder hardness. And there is a little difference in 7% GMB due to a decrease in complicated shear modulus values, which occur due to the high chemical content contained in GMB [4, 27, 29]. According to Shaban et al. (2017), the phase angle curves decrease with increasing polymer modifier concentration, which is caused by the polymer modifier binder reaching the failure temperature (46–64 °C). According to the G* (complex modulus) and (phase angles) values, 5% GMB produces better results, i.e., lower phase angle and higher complex modulus values. Polymer modifiers, according to studies, increase the storage stability and stiffness of the asphalt binder; adding 5% geopolymer to the asphalt yields the best results [4, 9, 12, 31].

7.4 Conclusions

In order to produce the pavement utilizing a geopolymer mix with asphalt concrete, the fly ash based geopolymer is used as a modifier in bitumen binder mixes in this chapter. The literature review indicates that geopolymer modified binder exhibits superior strength properties. In the fly ash based geopolymer bitumen binder, the findings of many investigations on the geopolymer mix are as follows: (GMB).

- The chemical composition of the precursor materials has an effect on the strength of the GMB mix; according to literature, it was noticed that the class F Fly ash produces better results when compared to the Class C Fly ash.
- According to the research, the physical properties of GMB are influenced by numerous geopolymer components, such as the L/S ratio and calcium content contained in precursor materials (Industrial by products) and the alkali activator ratio.

- It was discovered that 5% GMB had considerable physical characteristics, viscosity, and softening points. Geopolymers can also improve the workability of the bitumen binder and lower the mixing temperature.
- Rheological variables were altered by altering the percentage of the Geopolymer, although 5% GMB yielded acceptable values of G* 1 kPa (complex modulus) and (Phase angle).
- For sustainable development, the construction industry must increase the use of industrial byproducts as modifiers in asphalt mixtures. Fly ash is commonly utilised to create geopolymer materials. According to various studies, this fly ash-based geopolymer can also be employed with recyclable materials.

References

1. Erkus, Y., Kok, B. V., & Yilmaz, M. (2020). Evaluation of performance and productivity of bitumen modified by three different additives. *Construction and Building Materials, 261,* 120553. https://doi.org/10.1016/j.conbuildmat.2020.120553
2. Suresha, S. N., Varghese, G., Shankar, A. U. R. (2009). A comparative study on properties of porous friction course mixes with neat bitumen and modified binders. *Construction and Building Materials, 23,* 1211–1217. https://doi.org/10.1016/j.conbuildmat.2008.08.008.
3. Erkus, Y., Kok, B. V. (2022). Comparison of physical and rheological properties of calcium carbonate-polypropylene composite and SBS modified bitumen. *Construction and Building Materials.* https://doi.org/10.1016/j.conbuildmat.2022.130196.
4. Ali, S. I. A., Yahia, H. A. M., Ibrahim, A. N. H., & Mansob, R. A. A. (2017). *High temperatures performance investigation of geopolymer modified bitumen binders.* Taylor & Francis Group, London, ISBN 978-1-138-29595-7 Conference Paper, June 2017. https://doi.org/10.1201/978 1315100333-60.
5. Behnood, A., & Gharehveran, M. M. (2019). Morphology, rheology, and physical properties of polymer-modified asphalt binders. *European Polymer Journal, 112,* 766–791.
6. Nicholls, J. C. (1997). *Review of UK porous asphalt trials.* TRL Report 264, Transport Research Laboratory, London, United Kingdom.
7. ASTM D 3203. (2000). *Standard test method for percentage of air voids in compacted dense and open bituminous paving mixtures.*
8. Bardesi, A., et al. (1999). Use of modified bituminous binders, special bitumens and bitumens with additives in pavement applications. Technical Committee Flexible Roads (C8) World Road Association (PIARC).
9. Ali, S. I. A., Ismail, A., Yusoff, N. I. M., Karim, M. R., Al-Mansob, R. A., & Alhamali, D. I. (2015). Physical and rheological properties of acrylate–styrene–acrylonitrile modified asphalt cement. *Construction and Building Materials, 93,* 326–334.
10. Punith, V. S., Suresha S. N., Veeraragavan, A., Raju, S., & Bose, S. (2004). Characterization of polymer and fiber modified porous asphalt mixtures. In: *TRB 83rd annual meeting.* National Research Council.
11. Saride, S., & Jallu, M. (2020). Effect of fly ash geopolymer on layer coefficients of reclaimed asphalt pavement bases. *Journal of Transportation Engineering, Part B: Pavements, 146*(3), 04020033.
12. Ma, C.-K., Awang, A. Z., & Omar, W. (2018). Structural and material performance of geopolymer concrete: A review. *Construction and Building Materials, 186,* 90–102. https://doi.org/10.1016/j.conbuildmat.2018.07.111

13. Mullapudi, R. S., Bharath, G., & Reddy, N. G. (2021). Utilization of reclaimed asphalt pavement (RAP) material as a part of bituminous mixtures. In *Urban mining for waste management and resource recovery* (pp. 111–127). CRC Press.
14. Chien, Y. T., Hamid, R., & Kasmuri, M. (2012). Dynamic stress-strain behaviour of steel fiber reinforced high performance concrete with fly ash. *Advances in Civil Engineering., 2012*, 1–6. https://doi.org/10.1155/2012/907431
15. Tang, N., Deng, Z., Dai, J. G., Yang, K., Chen, C., & Wang, Q. (2018). Geopolymer as an additive of warm mix asphalt: Preparation and properties. *Journal of Cleaner Production, 192*, 906–915.
16. Chandra, K. S., Krishnaiah, S., Reddy, N. G., Hossiney, N., & Peng, L. (2021). Strength development of geopolymer composites made from red mud–fly ash as a subgrade material in road construction. *Journal of Hazardous, Toxic and Radioactive Waste, 25*(1), 04020068.
17. Bakare, M. D., Pai, R. R., Patel, S., & Shahu, J. T. (2019). Environmental sustainability by bulk utilization of fly ash and GBFS as road sub-base materials. *Journal of Hazardous, Toxic and Radioactive Waste, 23*, 4019011.
18. Komnitsas, K., & Zaharaki, D. (2007). Geopolymerization: A review and prospects for the minerals industry. *Minerals Engineering, 20* (14), 1261–1277.
19. Odion, D., Khattak, M. J., Abader, M., & Heim, N. (2019). Soil-geopolymer mixtures using recycled concrete aggregates for base and sub-base layers. In *Proceedings of the MATEC Web of Conferences*, Sibiu, Romania, 5–7 June 2019 (vol. 271, p. 2003). EDP Sciences.
20. Xu, H., & van Deventer, J. S. (2003). The effect of alkali metals on the formation of geopolymeric gels from alkali-feldspars. *Colloids and Surfaces A: Physicochemical and Engineering Aspects, 216*(1–3), 27–44.
21. Oderji, S. Y., Chen, B., Ahmad, M. R., & Shah, S. F. A. (2019). Fresh and hardened properties of one-part fly ash-based geopolymer binders cured at room temperature: Effect of slag and alkali activators. *Journal of Cleaner Production, 225*, 1–10.
22. He, J., Jie, Y., Zhang, J., Yu, Y., & Zhang, G. (2013). Synthesis and characterization of red mud and rice husk ash-based geopolymer composites. *Cement Concrete Compositions, 37*, 108–118. https://doi.org/10.1016/j.cemconcomp.2012.11.010.
23. Saoula, S., Mokhtar, K. A., Haddadi, H., & Ghorbel, E. (2008). Improvement of the stability of a modified bituminous binders within eva. *International Journal of Applied Engineering Research, 3*(4), 575–584.
24. Ibrahim, A. N. H., Yusoff, N. I. M., Akhir, N. M., & Borhan, M. N. (2016). Physical properties and storage stability of geopolymer modified asphalt binder. *Jurnal Teknologi, 78*(7–2), 133–138.
25. Al-Mansob, R. A., Ismail, A., Alduri, A. N., Azhari, C. H., Karim, M. R., & Yusoff, N. I. M. (2014). Physical and rheological properties of epoxidized natural rubber modified bitumens. *Construction and Building Materials, 63*, 242–248.
26. Memon, A. M., Sutanto, M. H., & Yusoff, N. I. M., et al. (2023). Rheological modeling and microstructural evaluation of oily sludge modified bitumen. *Case Studies in Construction Materials.* https://doi.org/10.1016/j.cscm.2023.e02039
27. Airey, G. D. (2003). Rheological properties of styrene butadiene styrene polymer modified road bitumens. *Fuel, 82*(14), 1709–1719.
28. Walubita, L. F., Alvarez, A. E., & Simate, G. S. (2011). Evaluating and comparing different methods and models for generating relaxation modulus master-curves for asphalt mixes. *Construction and Building Materials*, 2619–2626. https://doi.org/10.1016/j.conbuildmat.2010.12.010.
29. Lu, X., & Isacsson, U. (1997). Rheological characterization of styrene-butadiene-styrene copolymer modified bitumens. *Construction and Building Materials, 11*(1), 23–32.
30. Hamid, A., Alfaidi, H., Baaj, H., & El-Hakim, M. (2020). Evaluating fly ash-based geopolymers as a modifier for asphalt binders. *Advances in Materials Science and Engineering, 2020*, 1–11.
31. Golestani, B., Hyun, B., Moghadas, F., & Fallah, S. (2015). Nano clay application to asphalt concrete: Characterization of polymer and linear nano composite-modified asphalt binder and mixture. *Construction and Building Materials, 91*, 32–38.

Chapter 8
Mineral Carbonation of Mine Tailings for Long-Term Carbon Capture and Storage

Faradiella Mohd Kusin and Verma Loretta M. Molahid

8.1 Introduction

Climate change due to progressive carbon emission is vastly becoming a global concern. Carbon dioxide (CO_2) emission is considered the main contributor to global greenhouse gases that cause climate change. Thus, adoption of a reliable solution will help reduce the CO_2 emission for a long term. Carbon sequestration can be regarded as a feasible solution for the long-term CO_2 reduction in climate mitigation. This encompasses the strategies in reducing carbon emission such as focusing on low-carbon energy sources, developing renewable energies, applying geoengineering approach by enhancing the carbon sinks with reforestation and afforestation, and capturing and storing CO_2 which is known as carbon capture and storage (CCS). Following CCS, carbon capture utilization (CCU) has been introduced along with the realization that it is possible to gain something useful and valuable from the captured carbon.

 Carbon sequestration or CCS is derived from natural and man-made processes with the purpose of eliminating and diverting CO_2 from the atmosphere, which is then stored within the terrestrial environments, geological formations, or the ocean [1]. Generally, the sequestered CO_2 is kept permanently in a large storage site or reservoir to prevent its release back to the atmosphere [2]. Potential storage methods include CO_2 injection into deep ocean or underground geological structures and carbon fixation of industrial carbonates through mineral carbonation. The process

F. M. Kusin (✉) · V. L. M. Molahid
Department of Environment, Faculty of Forestry and Environment, Universiti Putra Malaysia, UPM, 43400 Serdang, Selangor, Malaysia
e-mail: faradiella@upm.edu.my

F. M. Kusin
Institute of Tropical Forestry and Forest Products (INTROP), Universiti Putra Malaysia, UPM, 43400 Serdang, Selangor, Malaysia

© The Author(s), under exclusive license to Springer Nature Singapore Pte Ltd. 2024   109
S. K. Das et al. (eds.), *Geoenvironmental and Geotechnical Issues of Coal Mine Overburden and Mine Tailings*, Springer Transactions in Civil and Environmental Engineering, https://doi.org/10.1007/978-981-99-6294-5_8

which is considered as part of CCU, offers a more stable and permanent solution in storing CO_2 where the risk of CO_2 leakage can be avoided [3]. The carbonate product formed can be used for developing new material or as supplementary material so that the use of natural minerals can be reduced [4]. Therefore, mineral carbonation can be regarded as an ideal approach in minimizing CO_2 emission in addition to creating a new economy, i.e., circularity, for CO_2 product utilization [5–7].

From an economic perspective, CCS technology may offer opportunities for the advancement of green technology and contribute to green economy. The captured CO_2 can be useful for other industries which can be used either directly or indirectly. In this regard, direct use of CO_2 is when it serves as a tool to ease a process, or to gain something useful with the help of CO_2. However, indirect use of CO_2 is when the end product of the captured carbon can be used to develop new materials or products.

This chapter discusses the fundamental processes of mineral carbonation technology and its potential adoption in the mining industry, and a range of feedstock materials for the process and potential utilization of mine tailings as feedstock for mineral carbonation application. Prospects and challenges in the adoption of this technology is also discussed.

8.2 Mineral Carbonation Technology

Mineral carbonation is a process of transforming gaseous CO_2 into stable carbonates upon reaction with minerals containing calcium or magnesium [3]. For instance, the reactions between CO_2 and calcium or magnesium oxides, which are derived from silicate minerals such as wollastonite, serpentine and olivine will form calcium and magnesium carbonates. Mineral carbonation technology can be considered as both CCS and CCU technologies, which depends on the process flow and the final location of the captured CO_2 or the carbonated product. The products of mineral carbonation (i.e., carbonates and silicates) are thermodynamically stable and it is safe to be disposed somewhere such as silicate mines (CCS) or reusing them for construction materials (CCU) [2]. It has been known that an advantage of using mineral carbonation technology adopting calcium and magnesium carbonates is due to the thermodynamically stable minerals that can withstand unexpected weather conditions of the storage site thus preventing CO_2 leakage to the atmosphere [8].

Feedstock materials for the mineral carbonation are largely available across the globe which provides a great opportunity for the adoption of the technology such as from industrial and mining waste, e.g., red gypsum, steel slag and silicate minerals. As the feedstock materials differ in nature, the process can be manipulated using different operating conditions and process variables to optimize the formation of carbonates [9].

One of the ideas for mineral carbonation implementation is that, it can be integrated with other industries that produce large amount of CO_2 emissions such as steel-making industry and mining industry, and at the same time, using their own waste to reduce the high CO_2 emission. This benefits not only the environment but also the industry where they can minimize their carbon footprint with minimal cost and gain additional revenue using the final product formed (carbonate product) from the carbonation process. As the integration of mineral carbonation in the industry is still considered as expensive, there is a need to evaluate the carbonate product formed so that it can be commercialized as a revenue-generating product which can be the stepping stone in promoting mineral carbonation technology [10].

8.2.1 Mineral Carbonation Methods

Mineral carbonation can be achieved through direct and indirect carbonation. The direct mineral carbonation deals with a single-step carbonation process, while indirect carbonation involves the extraction of ionic minerals from magnesium- and calcium-based minerals which will then undergo the carbonation step.

Direct Mineral Carbonation

Direct mineral carbonation encompasses two means of processes, i.e., gas–solid phase and aqueous phase. The direct gas–solid method is a relatively simple process however, it suffers from the slow reaction rate and is therefore not recommended for a big-scale project [11, 12]. Whereas the direct aqueous carbonation has been more commonly used in mineral sequestration compared to the direct gas–solid method [13–15]. Through the direct aqueous carbonation, the conversion efficiency can be enhanced by manipulating the operating variables under certain conditions. As for instance, 83.5% carbonation efficiency was obtained using wollastonite mineral ($CaSiO_3$) which is a calcium-bearing mineral, at a reaction temperature of 150 °C, reaction pressure of 40 bar and particle size < 30 μm [14]. In theory, an increase in temperature and pressure conditions, and decrease of particle size will increase the carbonation conversion efficiency. This has been proven in a study using red gypsum which is a high-calcium material that carbonation purity and efficiency can be improved by optimizing the operating parameters [13, 15].

Indirect Mineral Carbonation

Indirect mineral carbonation occurs through more than one step process prior to carbonation such as applying an initial extraction and precipitation. This method enables the dissolution and carbonation steps to be controlled separately [16]. The advantage of controlling these processes separately is that oxide minerals other than calcium and magnesium oxides can be removed to prevent interference during the carbonation process, which can improve the carbonation efficiency. This can be done by adding solvents or additives such as acids, bases or ammonium salts under certain conditions to leach out required cationic minerals and reacting it with carbon dioxide to form carbonate precipitation. Acids, bases and salt reagents, e.g., ammonium salts and ammonium chloride are commonly used in cation extraction. A cheaper alternative has been suggested using water as solvent in the dissolution of fly ash to facilitate cation leaching [17]. pH adjustment with the use of reagents can be adopted to provide feasible pH condition in enhancing carbonation efficiency [12, 18]. Furthermore, adopting low-reaction condition such as performing the mineral carbonation under ambient pressure and temperature can be a feasible approach to reduce energy consumption.

8.3   Potential Feedstock for Mineral Carbonation

Table 8.1 describes the conditions of the mineral carbonation process using Ca/ Mg/Fe-based minerals and waste materials in different operating conditions and the carbonation conversion efficiency obtained. There are a number of potential feedstocks for mineral carbonation which can be simplified into two types; natural silicate minerals and industrial waste. Both types contain essential minerals for carbonation such as Ca and Mg oxides, however, it varies in terms of its amount. Basically, natural silicate minerals have substantial amount of essential oxides for carbonation compared to industrial waste. Examples of commonly used natural silicate minerals are olivine, serpentine, wollastonite and limestone. For instance, in Table 8.1, wollastonite which is a Ca-based silicate mineral has a relatively high carbonation conversion of 91.1% [19]. Similarly, Yan et al. [14] also achieved a high carbonation conversion for wollastonite (83.5%) followed by serpentine (47.7%) and olivine (16.9%) from the same experiment. Furthermore, carbonation of limestone yielded 65% conversion efficiency [18]. Although the carbonation conversion was considered high, energy consumption used in the process was also high, suggesting that the use of these feedstocks has their uncertainty for real-life application.

On the other hand, the use of industrial waste provides an alternative for industries to reduce their waste in addition to a sustainable use of natural resources. According to Veetil and Hitch [20], the sources of the waste materials that have been utilized in the mineral carbonation process can be categorized into three namely, power plants with oil shale and coal as the fuel, construction and materials processing industry (i.e., cement, steel, aluminium and paper), and raw materials extraction site (i.e., natural minerals and ore deposit). Examples of commonly used waste from these sources are fly ash, steel slag, recycled concrete aggregate and mine waste. Referring to Table 8.1, carbon uptake capacity for fly ash is 0.11 g CO_2/kg in ambient condition [21]. Steel slag has been studied the most where a CO_2 capture capacity of 283 g CO_2/ kg was obtained [22] in higher operating conditions, while Revathy et al. [23] found an uptake capacity of 82 g CO_2/kg under lower operating conditions. Additionally, the use of recycled concrete aggregate is among the recent applications using waste material originated from construction and demolition waste. The use of this waste yielded 24.1% of carbonation conversion efficiency, and carbon uptake capacity of 27 g CO_2/kg of waste [24, 25]. For mine waste material, serpentinite-based tailings were used by Kemache et al. [26] and was able to obtain 0.22–0.25 g CO_2/kg of uptake capacity and 8.5–10.8% conversion efficiency. Similarly, Li and Hitch [10] achieved 8.7% of carbonation efficiency using mine waste rocks that contain Mg/ Fe-oxides.

Table 8.1 Summary of characteristics and operational conditions for mineral carbonation process

Feedstock material	Metal oxide (%)	Solution	Chemical treatment	Physical treatment	Heat treatment	Operating conditions	E_{CO2} (%)	CO_2 capture	Reference
Limestone	CaO = 98.52	Distilled water	–	Yes	Yes	P_{CO2} = 10 bar, T = 150 °C, particle size < 500 µm, stirring speed = 600 rpm	65	NA	[18]
Wollastonite Serpentine Olivine	W: CaO = 42.05 MgO = 1.21 S: CaO = 0.04 MgO = 40.59 O: CaO = 5.9 MgO = 32.27	0.6 M NaHCO$_3$	–	–	Yes	P_{CO2} = 40 bar, T = 150 °C, particle size < 30 µm, stirring speed = 600 rpm	W = 83.5 S = 47.7 O = 16.9	NA	[14]
Wollastonite	CaO = 46.06	Distilled water + 10% ammonia	Yes	Yes	–	P_{CO2} = 0.1 MPa, flow rate = 40 ml/min, T = 30 °C, time = 1 h, particle size < 20 µm, stirring speed = 600 rpm, flow rate = 40 ml/min	91.1	NA	[19]

(continued)

Table 8.1 (continued)

Feedstock material	Metal oxide (%)	Solution	Chemical treatment	Physical treatment	Heat treatment	Operating conditions	E_{CO2} (%)	CO_2 capture	Reference
Wollastonite	CaO = 36.28	Distilled water	–	Yes	–	P_{CO2} = 40 bar, T = 150 °C, time = 1 h, particle size < 30 μm, stirring speed = 1000 rpm	35.9	NA	[27]
Red Gypsum	CaO = 32.20 Fe$_2$O$_3$ = 28.99	Nanopure-demineralized water	–	Yes	–	P_{CO2} = 20 bar, T = 200 °C, time = 1 h, particle size < 45 μm, stirring speed = 1000 rpm	Ca = 12.53 Fe = 5.76	NA	[15]
Magnetite hematite + Iron	NA	–	–	Yes	–	P_{CO2} = 30 bar, T = 30 °C, time = 36 h, stirring speed = 400 rpm	NA	610 g CO$_2$/kg	[28]
Lizardite	MgO = 39.9	0.64 M NaHCO$_3$		Yes	Yes	P_{CO2} = 150 bar, T = 150 °C, time = 5 h, solid/liquid ratio = 15%	33	NA	[29]

(continued)

Table 8.1 (continued)

Feedstock material	Metal oxide (%)	Solution	Chemical treatment	Physical treatment	Heat treatment	Operating conditions	E_{CO2} (%)	CO_2 capture	Reference
Mine waste rock and Olivine	MW: MgO = 45.5 Fe_2O_3 = 10.4 O: MgO = 50.9 Fe_2O_3 = 8.4	1 M NaCl or 0.64 M $NaHCO_3$	–	Yes	–	P_{CO2} = 6 MPa, T = 185 °C, time = 1 h, particle size < 106 μm, stirring speed = 1500 rpm, liquid/solid ratio = 1	MW = 8.7 O = 5.2	NA	[10]
Serpentinite-based tailings	R: MgO = 46.3, Fe_2O_3 = 7.6 P: MgO = 44.3, Fe_2O_3 = 12.1	Distilled water	–	Yes	Yes	P_{CO2} = 10 atm, T = 32–40 °C, time = 6 h, particle size = 67 μm, P_{CO2} = 8 atm, T = 32–40 °C, time = 6 h, particle size = 47 μm, solid/liquid ratio = 150 g/L, gas/liquid ratio = 3, stirring speed = 600 rpm	R = 8.5 P = 10.8	R = 0.22 g CO_2/g P = 0.25 g CO_2/g	[26]

(continued)

Table 8.1 (continued)

Feedstock material	Metal oxide (%)	Solution	Chemical treatment	Physical treatment	Heat treatment	Operating conditions	E_{CO2} (%)	CO_2 capture	Reference
Ultramafic nickel ores	MgO = 45.8	Distilled water 1 M NaCl + 0.64 M NaHCO$_3$		Yes	Yes	P_{CO2} = 12.4 MPa, T = 155 °C, time = 1 h, particle size = μm, solid/liquid ratio = 20%, stirring speed = 1000 rpm	36.6	18.3 g CO$_2$/100 g	[7]
Steel slag	CaO = 33.19, Fe$_2$O$_3$ = 38.19	Distilled water	–	Yes	–	P_{CO2} = 10.6 bar, T = ambient, time = 10 min, particle size < 67 μm, liquid/solid ratio = 10, gas/liquid ratio = 3, stirring speed = 600 rpm	NA	0.05 g CO$_2$/g	[30]
EAF Steel slag	CaO = 28.27, Fe$_2$O$_3$ = 24.25	Distilled water	–	Yes	–	P_{CO2} = 6 bar, T = 30 °C, time = 3 h, particle size < 2 mm, stirring speed = 500 rpm, liquid/solid ratio = 1	27	82 g CO$_2$/kg	[23]

(continued)

Table 8.1 (continued)

Feedstock material	Metal oxide (%)	Solution	Chemical treatment	Physical treatment	Heat treatment	Operating conditions	E_{CO2} (%)	CO_2 capture	Reference
Steel slag (Blended hydraulic slag cement)	CaO = 52.82	Distilled water	–	Yes	–	P_{CO2} = 48.26 bar, T = 160 °C, time = 12 h, particle size <44 μm, liquid/solid ratio = 10 mL/g	68.3	283 g CO_2/kg	[22]
Steel slag	NA	Distilled water		Yes		P_{CO2} = 3 bar, T = 30 °C, time = 2 h, particle size < 0.105 mm, liquid/solid ratio = 0.4	13	130 g CO_2/kg	[31]
Steel slag	CaO = 31.7 FeO = 35.5	Nanopure-demineralized water	–	Yes	–	P_{CO2} = 19 bar, T = 100 °C, time = 0.5 h, particle size = 38 μm, stirring speed = 500 rpm, liquid/solid ratio = 2 kg/kg	74	185 g CO_2/kg	[11]
Recycled concrete aggregates	CaO = 63.62 FeO = 3.08 MgO = 1.46	–	–	–	–	Flow rate = 5 L/min, T = 25 °C, relative humidity = 50, time = 168 h	24.1	NA	[24]

(continued)

Table 8.1 (continued)

Feedstock material	Metal oxide (%)	Solution	Chemical treatment	Physical treatment	Heat treatment	Operating conditions	E_{CO2} (%)	CO_2 capture	Reference
Recycled masonry aggregate	CaO = 12.8 FeO = 0.74 MgO = 1.65	–	–	Yes	–	Time = 7 d	0.55	5.55 g CO_2/kg	[25]

(continued)

Table 8.1 (continued)

Feedstock material	Metal oxide (%)	Solution	Chemical treatment	Physical treatment	Heat treatment	Operating conditions	E_{CO2} (%)	CO_2 capture	Reference
Carbide slag	CS: CaO = 69.19 MgO = 1.10	Distilled water + 1.67 M Na$_2$SiO$_3$ + 0.84 M NH$_4$Cl	–	Yes	–	P$_{CO2}$ = 0.6 MPa, T = 55 °C, time = 125 min, particle size = 74–150 µm, liquid/solid ratio = 6	CS = 64 WL = 72 SS = 52 WC = 32	CS = 0.44 kg/ kg WL = 0.58 kg/kg SS = 0.20 kg/ kg WC = 0.07 kg/kg	[32]
Waste lime	WL: CaO = 81.32 FeO = 0.0.34								
Steel slag	SS: CaO = 38.91 Fe$_2$O$_3$ = 23.11FeO = 21.62 MgO = 7.61								
Waste concrete	WC: CaO = 23.51 Fe$_2$O$_3$ = 1.12 FeO = 0.72 MgO = 1.64								

(continued)

Table 8.1 (continued)

Feedstock material	Metal oxide (%)	Solution	Chemical treatment	Physical treatment	Heat treatment	Operating conditions	E_{CO2} (%)	CO_2 capture	Reference
Coal fly ash	$CaO = 7.2$ $Fe_2O_3 = 7.8$	Deionized water	Yes	–	–	P_{CO2} = ambient pressure, flow rate = 2 ml/min, T = ambient temperature (25 °C), stirring speed = 700 rpm solid/liquid ratio = 100 g/L	NA	0.11 g CO_2/g	[21]
Lignite fly ash	NA	Water	–	–	–	P_{CO2} = 0.15 atm, T = 40–80 °C, time = 2 h, particle size < 200 μm, stirring speed = 1500 rpm, liquid/solid ratio = 0.12 g/g	53	4.8 mmol g^{-1}	[33]

Note P_{CO2}—CO_2 pressure; E_{CO2}—Carbonation efficiency; T—Temperature; Chemical treatment—use of additives or leaching agents (e.g. acids or bases); Physical treatment—particle size alteration (e.g., crushing or grinding); Heat treatment—variation of temperature at high degree temperatures

Carbonation using waste materials can be considered noteworthy considering the limitation on an essential oxide amount. By carefully considering the utilization of these materials, waste accumulation can be reduced, as well as its adverse impact on the surrounding environment. However, more efforts need to be put into investigating the specific waste materials used for mineral carbonation as the composition varies widely in terms of mineralogical and chemical components.

8.4 Mine Waste Utilization for Carbon Sequestration

The increase in demand for raw minerals such as gold, iron, coal, chrome, copper, fluorspar, zinc, diamond and manganese towards modernization has made waste generation from mining industry inevitable [34]. It is of no doubt that mining companies will continue to face high accumulation of waste. Additionally, there have been health and environmental issues associated with mining activities. For example, weathering of the mine waste can cause contamination of surface water with heavy metals, causing the water to become acidic mine drainage [35]. On the other hand, mining and mineral extraction, ore processing, and post-mining activities such as handling of mine tailings can cause health effects to humans upon exposure for a long time [34]. Furthermore, accumulation of mine waste will eventually occupy all available land for storage which can be a burden as the location needs to be rehabilitated when the mining operation ends [36].

Despite this, mining waste can be utilized as a feedstock material for carbon capture in the mining industry. Mineral carbonation technology provides an opportunity for mining waste utilization in carbon sequestration to permanently store CO_2 in stable carbonates. Mining waste such as waste rocks, soils and mine tailings enable long-term CO_2 storage because of the mineral availability and reactivity with the CO_2. The overall concept of mineral carbonation utilizing mine waste is as illustrated in Fig. 8.1. Targeted minerals have often been the best approach to obtain minerals that can react with CO_2 to optimize the conversion rate and products. However, this chapter deals mostly with mining waste materials produced as waste rocks, tailings and residues from mining operations rather than purposely mine for targeted minerals. These materials are vastly produced as a consequence of mining activities and have often been left abandoned. This could bring environmentally undesirable impacts on our ecosystems and on earth generally. When mine tailings are removed from the ground in the mining process and exposed to the atmosphere and to water, they can react to form new minerals that trap CO_2 from the atmosphere.

Clearly, mine tailings utilization in mineral carbonation serves as an alternative approach to minimize damages or problems related to mining waste. Apart from reducing the adverse impact on environment, the mining industry can also reduce the expenditure required in handling mine waste, as well as managing the mine site for rehabilitation. In addition, there is an opportunity for the mining industry to

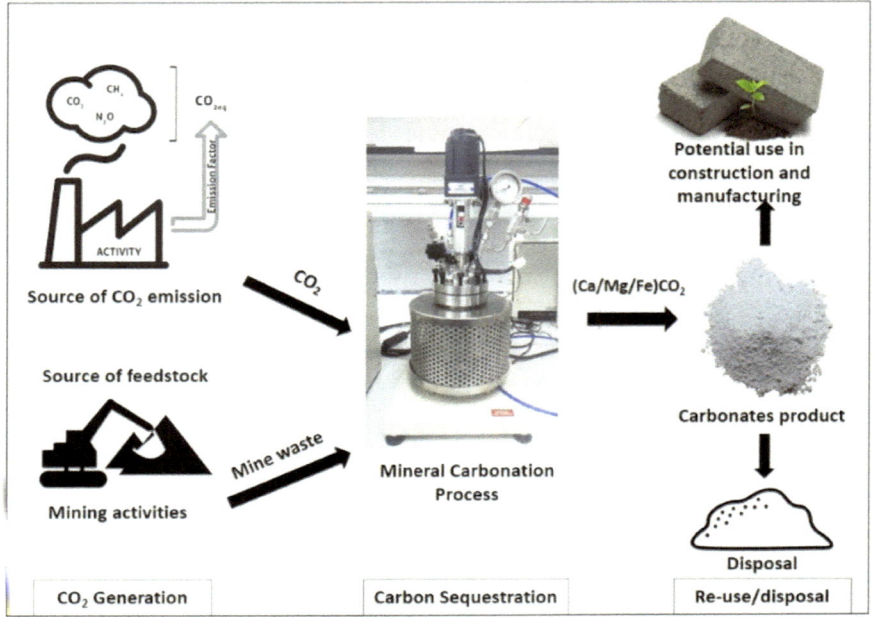

Fig. 8.1 Conceptual mineral carbonation process for carbon sequestration of mining waste

earn additional revenue when integrating mineral carbonation in producing value-added carbonated products [36]. In return, it can be commercialized as a carbon-negative product, where CO_2 is permanently captured. Carbonated product can also be used in the construction sector such as partial replacement for cement or aggregate in concrete- or cement-based materials [4]. Although it is still at early progress, the benefits from this approach are noteworthy and will be a great step towards low-carbon future.

8.5 Influencing Factors for Mineral Carbonation

In the mineral carbonation process, major reactions that occur through aqueous carbonation are greatly dependent on the operating conditions applied. The major reactions include CO_2 dissolution in slurry, metal ion liberation from its minerals (e.g. Mg^{2+}, Ca^{2+}, Fe^{3+}) and carbonate product precipitation which involves the reaction between the metal ions and bicarbonate ions (HCO_3^-) [33]. These major reactions are governed by the operating conditions which are crucial in achieving optimum carbonate yield throughout the process. There are a number of factors that are influential for mineral carbonation. Some of the studied factors are reaction temperature, particle size, CO_2 pressure, pH, reaction time, water content and stirring rate. However, details on reaction temperature, CO_2 pressure, particle size, and pH condition are discussed below as these are the main parameters investigated in most studies.

Reaction Temperature

Increase in reaction temperature has two differing effects on the carbonation reaction; faster metal ion leaching from its mineral and lower CO_2 solubility in the solution [11]. This was found that the carbonation rate increases with temperature and when it reaches higher than 175 °C, the degree of carbonation started to decline. This shows that low CO_2 solubility limits carbonate precipitation at elevated temperatures. Yan et al. [14] reported similar increasing trends from 80–150 °C for wollastonite, serpentine and olivine. However, as temperature increases to 200 °C, carbonation decreases slightly for serpentine, while it increases to nearly 60% for wollastonite and to 9.3% for olivine. The optimum temperature was found to be 185 °C for all these minerals. Azdarpour et al. [13] found a declined carbonation efficiency at a temperature of 200 °C despite increasing carbonate purity with increasing temperature. Ukwattage et al. [9] experienced a decrease in carbonation efficiency at temperature > 60 °C because of low CO_2 solubility. Fagerlund et al. [37] obtained low final carbonation rate due to over-increase in temperature. A study has reported an optimum temperature for wollastonite carbonation which was at 150 °C, while olivine was at 185 °C [10]. Apart from wollastonite and olivine, the optimum temperature for limestone was 150 °C [18]. In addition, the optimum temperature when utilizing industrial waste as the carbonation agent was reported to be 25–90 °C for steel slag [9, 23, 32, 38], 32–40 °C for serpentinite-based tailings [26], 25–80 °C for fly ash [21, 32], and 55°C for carbide slag, waste lime and waste concrete [32]. Therefore, it is important to use an appropriate temperature for specific feedstock so that an optimum carbonation rate can be achieved or the process can be accelerated.

Carbon Dioxide Pressure

Another important parameter that can enhance mineral carbonation is CO_2 pressure. Higher partial CO_2 pressure causes more CO_2 solubility in aqueous media, forming carbonic acid, thus increasing the formation of bicarbonate ions [39]. Therefore, more bicarbonates can react with the metal ions. Previous studies have revealed a direct relationship between carbonate purity and efficiency with CO_2 pressure, where increased CO_2 pressure improves both carbonate purity and efficiency [13, 15]. The study applied CO_2 pressure ranging from 1 to 70 bar and found 100% of carbonation efficiency and 98% of $CaCO_3$ purity which was achieved at CO_2 pressure of 8 bar for direct aqueous carbonation using red gypsum [13]. At 70 bar CO_2 pressure, the maximum $FeCO_3$ and $CaCO_3$ in the final products of red gypsum were 5.01% and 14.28%, respectively [15]. Another research was using basic-oxygen-furnace slags as a feedstock for mineral carbonation and had found that the maximum carbonation conversion was 93.5% at 1 bar CO_2 pressure and 65 °C [22]. Notwithstanding this, a higher CO_2 pressure which is between 100 and 150 bar will be able to increase the carbonation rate and reduce the reaction time [3]. Thus, in sequestrating more

CO_2, a higher CO_2 pressure seems to have an important influence on the formation of carbonate, however the application is limited by cost and energy constraints.

Particle Size

Generally, smaller particle size will increase the carbonation degree. For example, a carbonation rate of 91.1% was achieved using particle size of 20 μm [19], while similarly a high carbonation was achieved using feedstock with particle size of between 25 and 37 μm [27, 38]. Additionally, the following carbonation rate was obtained using particle size of < 45 μm for red gypsum (10.8%), < 47 μm for serpentinite-waste tailings (12.53%), < 10 μm for steel slag (13%), and 2 mm steel slag (27%), respectively [9, 15, 23, 26, 31]. Smaller particle size will facilitate CO_2 gas exchange with the entire mass of the waste [25]. Decreasing the particle size provides more available surface area for the reaction to occur as carbonation is proportional with the total specific surface area of a particle, thus a higher carbonation rate can be expected. A smaller particle size will also improve the metal ion leaching rate from its matrix up to 99% [40]. Leaching is an important step for the indirect carbonation process where the leached metal ions will be able to react with bicarbonate ions more effectively prior to form carbonates. Therefore, it is important that the right particle size be used in the carbonation process which can benefit both the leaching and precipitation processes.

pH Condition

Another controlling factor that can affect the carbonation rate is pH condition. The pH condition will determine the availability of metal ions essential for carbonate precipitation to occur [3]. Magbitang and Lamorena [41] reported that altering pH (i.e., pH 4–10) can improve metal ion dissolution and CO_2 reaction in an aqueous solution. However, pH condition may have contradicting effect on the carbonation process. According to Wang et al. [42], the carbonation rate can be increased by increasing the concentration of Ca/Mg/Fe ions or carbonate ions. It appears that low pH can aid Ca/Mg/Fe extraction from the matrix which increases its availability for CO_2 gas to react with, however, it also lowers the carbonate ion concentration in the solution [7, 13]. Increase in pH results in higher concentration of carbonate ions, but lowers the availability of Ca/Mg/Fe ions. Some studies suggested that the optimum pH for carbonation to occur is greater than 10 [43, 44]. In addition, some suggested adding sodium bicarbonate, $NaHCO_3$ as a means to increase carbonate ion concentration and successfully achieved 70% conversion rate in 2 h [45]. $NaHCO_3$ and sodium chloride, $NaCl$ are common additives used as buffering agents to maintain the solution pH in the range of 7–8 [10, 46]. In essence, pH needs to be carefully controlled in order to balance the opposite effect it gives to the major reactions in the carbonation process. In other words, the chosen pH needs to enhance the metal ions extraction, while at the same time able to facilitate the precipitation process by providing more carbonate ions.

On a general note, mineral carbonation is a relatively slow process if it occurs naturally, however, the reactions can be accelerated by adjusting the operating variables. If the process was accelerated, large mines may have the capacity to sequester more CO_2, thereby providing a great potential to offset greenhouse gas emissions from the mining industry. While it has been known that alkaline materials are often preferred for the process, further understanding of the changing pH on the process can be evaluated. Similarly, other variables such as small-sized particles could be an option for a better carbonation reaction. Varying the parameters will help identify the variables that can be applied to optimize the carbonation conversion. This is particularly important to identify the limits of conversion rate and assess process efficiency. Understanding the process efficiency will provide insights into large-scale application as carbonation efficiency can be estimated and compared for different feedstock materials at different operating variables.

8.6 Case Study: Carbon Sequestration of Mine Tailings

From a Malaysian experience, a range of mining waste have been studied for their potential as sequestering agents through mineral carbonation including those that are generated from metal mines such as iron ore, gold, tin and bauxite and non-metal mines such as limestone. Figure 8.2 shows the abundance of mine tailings that are generated from limestone, iron ore and gold mining, respectively.

Mine waste is regarded as a potential feedstock for mineral carbonation due to its alkaline nature and affinity to react with CO_2, a mild acid. Potential feedstocks

Fig. 8.2 Mine tailings **a** stockpile of limestone waste **b** waste dump of iron mine **c** waste dump of gold mine **d** stockpile of gold mine waste

for carbon sequestration from mining waste consist of alkaline earth metal-bearing silicates, hydroxide minerals and silicate waste rocks or tailings that are rich in divalent cations such as magnesium, calcium and iron oxide. Mineralogical and chemical composition play important roles in determining the suitability of these materials for mineral carbonation. Silicate minerals that are present in mine waste act as an important agent for trapping CO_2 and store it as stable carbonates. Ca and Mg silicates have been known to have a great capacity for sequestering CO_2 and are particularly useful for mineral carbonation. Table 8.2 demonstrates the types of minerals that are available from respective mines, which are potential minerals to undergo mineral carbonation. Limestone waste particularly contain Ca–Mg-silicates of akermanite and carbonate minerals of calcite and dolomite. Iron mine waste consists of Ca-silicates of anorthite and wollastonite and Fe-oxides of magnetite and hematite, and Ca–Mg-silicates of diopside. However, the gold mine waste comprises of Mg-silicates of sepiolite and chlorite-serpentine and Fe-silicates of stilpnomelane. Upon reaction with CO_2, these minerals will be able to capture the gas and convert it into a stable carbonate form.

In view of its potential as feedstock for the mineral carbonation process, carbonation efficiency can be determined from the utilization of mine tailings as summarized in Table 8.3. Varying parameters and operating conditions were taken into account during the lab-scale experiment including particle size, reaction temperature, pH and CO_2 pressure. Findings indicated that small-sized particles of < 38 μm, reaction temperature of 80–200 °C and pH 10–12 were able to achieve a feasible carbonation conversion, which was conducted at an ambient CO_2 pressure (1 bar) [47, 49].

Therefore, it has been shown that these types of mine tailings are potential candidates for mineral carbonation, which is governed by the availability of Ca/Mg/Fe-bearing minerals and their oxide composition. Clearly, reutilizing mine waste may give an opportunity for restoring natural resources while tackling the issues of carbon emission at the mining site in general.

Table 8.2 Potential minerals for mineral carbonation of mine tailings	Mining waste type	Potential mineral	Reference
	Limestone	Akermanite	[4]
		Calcite	
		Dolomite	
	Iron ore	Anorthite	[47]
		Wollastonite	
		Magnetite	
		Hematite	
		Diopsite	
		Siderite	
	Gold	Sepiolite	[48]
		Chlorite-serpentine	
		Stilpnomelane	

Table 8.3 Carbonation conversion efficiency of mine tailings

Waste material	Carbonation efficiency, E_{CO2}				Reference
Iron mine tailings	Particle size	Temperature	pH	CO_2 pressure	[47]
	<38 μm E_{CO2}: 6.66%	80°C E_{CO2}: 2.68%	pH 8 E_{CO2}: 2.41%	1 bar	
	<75 μm E_{CO2}: 2.78%	200°C E_{CO2}: 5.82%	pH 12 E_{CO2}: 5.85%		
Limestone mine waste	Particle size	Temperature	pH	CO_2 pressure	[49]
	<38 μm	80 °C	pH 10	1 bar	
	E_{CO2}: 7.53%				

8.7   Challenges in the Adoption of Mineral Carbonation for Mining Waste

Among the major challenges in the adoption of mineral carbonation are the concerns about cost and energy consumption [50]. Moreover, the limitation of the current scale-up applications of mineral carbonation are mainly due to large amount of costs and energy usage. Generally, the costs and energy consumption for mineral carbonation are associated with the mineral carbonation plant, pre-treatment requirements, operating conditions, use of additives and disposal of the reaction products [3]. As mining is often associated with the issues of CO_2 emission and waste generation, mineral carbonation can be an option to tackle both issues if operated under feasible conditions, such as considering the reduction of costs and energy. However, another challenge for mineral carbonation is to increase the carbonation reaction to be feasible for large-scale deployment. It has been known that the slow dissolution kinetics and the potentially high energy required are the main challenges to manipulate mineral processing. An economic analysis of mineral carbonation adoption for mining industry has been evaluated in an earlier study that found an operation cost of USD82.5 per t CO_2 [51]. Cost comparisons from other industrial waste and natural minerals could be the benchmark for mineral carbonation adoption with mine waste in the long term. Generally, the cost of indirect mineral carbonation is estimated to be higher than the direct mineral carbonation which accounts for about 40% of the total cost [52]. Despite this, opportunity for mineral carbonation with mining waste can be seen through the integration of the waste material to reduce the environmental impact, and utilization potential of the captured carbon into revenue-generating products to counterbalance the energy-cost implications.

8.8   Current State-of-The-Art of Mineral Carbonation Technology

Industrial-scale mineral carbonation has taken place in some instances global wide. For example, CarbonCure Technologies (Canada), Mineral Carbonation International (Australia), Carbon8 Systems (United Kingdom), and Solidia Technologies (United States) are the companies that are mainly focusing on construction materials production through the mineral carbonation process. Besides having it successfully implemented at industrial-scale, the concept of this technology has been widely promoted to encourage more research and development at various levels [50]. Although it is still in early stage, this shows that the nation is well-exposed to the benefit of this technology.

Awareness on mineral carbonation technology has been greater, and this is particularly true mostly in places where the technology has reached commercialization stage. The success of CarbonCure technology, a Canadian cleantech company has taken the interest of many individuals especially the local communities in adapting to climate change mitigation actions, while at the same time generating income. Starting in 2007, CarbonCure have nearly 200 companies across North America and Southeast Asia that have adopted their technology. To date, more than 5 million cubic yards of concrete made with recycled carbon dioxide have been delivered to the construction project across the globe. Pan-United Corporation Ltd. Singapore is the first company in Asia that has adopted CarbonCure technology, initiating an environmentally friendly scenario and low-carbon industry in Singapore [51].

8.9   Conclusions

From a mining perspective, utilization of mine waste or mine tailings would be advantageous in restoring natural mineral resources. Mine tailings that have often been associated with undesirable impacts on the environment can be seen as a valuable resource for sequestering carbon dioxide. While they are abundantly generated from mining operations, mine tailings may serve as reservoir or sinks for atmospheric carbon sequestration at mining sites. Depending on the availability of reactive minerals and input of carbon dioxide, the process may form part of a sustainable mining in the future. Despite current challenges in terms of implementation costs, on-site deployment method and technological acceptance by the key industry players, this technology remains a reliable option for future sustainability in the mining sector. This is also in line with the concept of circular economy such that mine waste can be turned into a useful resource. Therefore it can be learned that the technological approach would likely be a reliable solution in response to global carbon reduction and climate mitigation for the mining industry.

References

1. United State of Geological Survey (USGS). Carbon Sequestration to Mitigate Climate Change. Retrieved June 24, 2018, from https://pubs.usgs.gov/fs/2008/3097/pdf/CarbonFS.pdf#page=1&zoom=auto,-99,798.
2. IPCC. (2005). Carbon dioxide capture and storage. In B. Metz, O. Davidson, H. D. Coninck, M. Loos, & L. Meyer (Eds.), *IPCC special report prepared by working group III of the intergovernmental panel on climate change* (442 pp.). Cambridge University Press.
3. Sanna, A., Uibu, M., Caramanna, G., Kuusik, R., & Maroto-Valer, M. M. (2014). A review of mineral carbonation technologies to sequester CO_2. *Chemical Society Reviews, 43*(23), 8049–8080.
4. Kusin, F. M., Hasan, S. N. M. S., Hassim, M. A., & Molahid, V. L. M. (2020). Mineral carbonation of sedimentary mine waste for carbon sequestration and potential reutilization as cementitious material. *Environmental Science and Pollution Research, 27*, 12767–12780.
5. Lackner, K. S., Wendt, C. H., Butt, D. P., Joyce, E. L., & Sharp, D. H. (1995). Carbon dioxide disposal in carbonate minerals. *Energy, 20*(11), 1153–1170.
6. Jacobs, A. D., & Hitch, M. (2011). Experimental mineral carbonation: Approaches to accelerate CO_2 sequestration in mine waste materials. *International Journal of Mining, Reclamation and Environment, 25*(4), 321–331.
7. Bobicki, E. R., Liu, Q., Xu, Z., & Zeng, H. (2012). Carbon capture and storage using alkaline industrial wastes. *Progress in Energy and Combustion Science, 38*(2), 302–320.
8. Teir, S., Eloneva, S., Fogelholm, C. J., & Zevenhoven, R. (2006). Stability of calcium carbonate and magnesium carbonate in rainwater and nitric acid solutions. *Energy Conversion and Management, 47*(18–19), 3059–3068.
9. Ukwattage, N. L., Ranjith, P. G., & Li, X. (2017). Steel-making slag for mineral sequestration of carbon dioxide by accelerated carbonation. *Journal of the International Measurement Confederation, 97*, 15–22.
10. Li, J., & Hitch, M. (2017). A review on integrated mineral carbonation process in ultramafic mine deposit. *Geo-Resources Environment and Engineering*, 148–154.
11. Huijgen, W. J. J., & Comans, R. N. J. (2005). Carbon dioxide sequestration by mineral carbonation: Literature review update 2003–2004, ECN-C—05-022, Energy Research Centre of The Netherlands, Petten, The Netherlands.
12. Sipilä, J., Teir, S., & Zevenhoven, R. Carbon dioxide sequestration by mineral carbonation literature review update 2005–2007. Retrieved June 16, 2018, from https://remineralize.org/wp-content/uploads/2015/10/LITR1.pdf.
13. Azdarpour, A., Asadullah, M., Junin, R., Manan, M., Hamidi, H., & Mohammadian, E. (2014). Direct carbonation of red gypsum to produce solid carbonates. *Fuel Processing Technology, 126*, 429–434.
14. Yan, H., Zhang, J., Zhao, Y., Liu, R., & Zheng, C. (2015). CO_2 Sequestration by direct aqueous mineral carbonation under low-medium pressure conditions. *Journal of Chemical Engineering of Japan, 48*(11), 937–946.
15. Azdarpour, A., Karaei, M. A., Hamidi, H., Mohammadian, E., & Honarvar, B. (2018). CO_2 sequestration through direct aqueous mineral carbonation of red gypsum. *Petroleum*, 1–10.
16. Jacobs, A. D. (2014). *Quantifying the mineral carbonation potential of mine waste mineral: A new parameter for geospatial estimation* (Doctor of Philosophy Thesis). Faculty of Graduate and Postdoctoral Studies, (Mining Engineering), University of Columbia, Vancouver, 232.
17. Yogo, K., Eikou, T., & Tateaki, Y. (2005). Method for fixing carbon dioxide, Patent, JP2005097072, 14.4.2005.
18. Han, D. R., Namkung, H., Lee, H. M., Huh, D. G., & Kim, H. T. (2015). CO_2 sequestration by aqueous mineral carbonation of limestone in a supercritical reactor. *Journal of Industrial and Engineering Chemistry, 21*, 792–796.
19. Ding, W., Fu, L., Ouyang, J., & Yang, H. (2014). CO_2 mineral sequestration by wollastonite carbonation. *Physics and Chemistry of Minerals, 41*, 489–496.

20. Veetil, S. P., & Hitch, M. (2020). Recent developments and challenges of aqueous mineral carbonation: A review. *International Journal of Environmental Science and Technology, 17*(10), 4359–4380.

21. Jo, H. Y., Kim, J. H., Lee, Y. J., Lee, M., & Choh, S. J. (2012). Evaluation of factors affecting mineral carbonation of CO_2 using coal fly ash in aqueous solutions under ambient conditions. *Chemical Engineering Journal, 183*, 77–87.

22. Chang, E. E., Pan, S. Y., Chen, Y. H., Chu, H. W., Wang, C. F., & Chiang, P. C. (2011). CO_2 sequestration by carbonation of steelmaking slags in an autoclave reactor. *Journal of Hazardous Materials, 195*, 107–114.

23. Revathy, T. D. R., Palanivelu, K., & Ramachandran, A. (2016). Direct mineral carbonation of steelmaking slag for CO_2 sequestration at room temperature. *Environmental Science and Pollution Research, 23*(8), 7349–7359.

24. Pu, Y., Li, L., Wang, Q., Shi, X., Fu, L., Zhang, G., Luan, C., & Abomohrab, A. E. F. (2020). Accelerated carbonation treatment of recycled concrete aggregates using flue gas: A comparative study towards performance improvement. *Journal of CO2 Utilization, 43*, 101362.

25. Suescum-Morales, D., Kalinowska-Wichrowska, K., Fernández, J. M., & Jiménez, J. R. (2021). Accelerated carbonation of fresh cement-based products containing recycled masonry aggregates for CO2 sequestration. *Journal of CO2 Utilization, 46*.

26. Kemache, N., Pasquier, L. C., Cecchi, E., Mouedhen, I., Blais, J. F., & Mercier, G. (2017). Aqueous mineral carbonation for CO_2 sequestration: From laboratory to pilot scale. *Fuel Processing Technology, 166*, 209–216.

27. Yan, H., Zhang, J., Zhao, Y., & Zheng, C. (2013). CO_2 sequestration from flue gas by direct aqueous mineral carbonation of wollastonite. *Science China Technological Sciences, 56*(9), 2219–2227.

28. Mendoza, E. Y. M., Santos, A. S., López, E. V., Drozd, V., Durygin, A., Chen, J., & Saxena, S. K. (2019). Siderite formation by mechanochemical and high pressure-high temperature processes for CO_2 capture using iron ore as the initial sorbent. *Processes, 7*(735).

29. Benhelal, E., Rashid, M. I., Rayson, M. S., Prigge, J. D., Molloy, S., Brent, G. F., Cote, A., Stockenhuber, M., & Kennedy, E. M. (2018). Study on mineral carbonation of heat activated lizardite at pilot and laboratory scale. *Journal of CO2 Utilization, 26,* 230–238.

30. Ghacham, A. B., Pasquier, L. C., Cecchi, E., Blais, J. F., & Mercier, G. (2016). CO_2 sequestration by mineral carbonation of steel slags under ambient temperature: Parameters influence, and optimization. *Environmental Science and Pollution Research, 23*, 17635–17646.

31. Baciocchi, R., Costa, G., Polettini, A., & Pomi, R. (2009). Influence of particle size on the carbonation of stainless steel slag for CO_2 storage. *Energy Procedia, 1*(1), 4859–4866.

32. Chang, J., Fang, Y., & Wang, J. (2015). CO_2 sequestration by solid residues carbonation for building materials preparation. *International Journal of Advances in Agricultural and Environmental Engineering, 2*(1), 49–53.

33. Bauer, M., Gassen, N., Stanjek, H., & Peiffer, S. (2011). Carbonation of lignite fly ash at ambient T and P in a semi-dry reaction system for CO_2 sequestration. *Applied Geochemistry, 26*(8), 1502–1512.

34. Vallero, D.A., & Blight, G. (2019). Mine waste: A brief overview of origins, quantities, and methods of storage. In *Waste Streams (and Their Treatment)* (2nd ed., pp. 129–151). Elsevier Inc.

35. Affandi, F. N. A., Kusin, F. M., Sulong, N. A., & Madzin, Z. (2018). Hydrogeochemical assessment of mine-impacted water and sediment of iron ore mining. *IOP Conference Series Earth and Environmental Science, 140*, 012–023.

36. Hitch, M., Ballantyne, S. M., & Hindle, S. R. (2010). Revaluing mine waste rock for carbon capture and storage. *International Journal of Mining, Reclamation and Environment, 24*, 64–79.

37. Fagerlund, J., Nduagu, E., Romão, I., & Zevenhoven, R. (2012). CO_2 fixation using magnesium silicate minerals part 1: Process description and performance. *Energy, 41*(1), 184–191.

38. Yadav, S., & Mehra, A. (2017). Experimental study of dissolution of minerals and CO_2 sequestration in steel slag. *Waste Management, 64*, 348–357.

39. Harrison, A. L., Power, I. M., & Dipple, G. M. (2013). Accelerated carbonation of brucite in mine tailings for carbon sequestration. *Environmental Science & Technology, 47*, 126–134.
40. Rahmani, O., Highfield, J., Junin, R., Tyrer, M., & Pour, A. B. (2016). Experimental investigation and simplistic geochemical modeling of CO_2 mineral carbonation using the mount Tawai peridotite. *Molecules, 21*(3).
41. Magbitang, R. A., & Lamorena, R. B. (2016). Carbonate formation on ophiolitic rocks at different pH, salinity and particle size conditions in CO_2-sparged suspensions. *International Journal of Industrial Chemistry, 7*(4), 359–367.
42. Wang, A., Zhong, D., Zhu, H., Guo, L., Jiang, Y., Yang, X., & Xie, R. (2019). Diagentic features of illite in Upper Triassic Chang-7 tight oil sandstones, Ordos Basin. *Geoscience Journal, 23*, 281–298.
43. Olajire, A. A. (2013). A review of mineral carbonation technology in sequestration of CO_2. *Journal of Petroleum Science and Engineering, 109*, 364–392.
44. Park, A. H. A., & Fan, L. S. (2004). CO_2 mineral sequestration: Physically activated dissolution of serpentine and pH swing process. *Chemical Engineering Science, 59*(22–23), 5241–5247.
45. Eikeland, E., Blichfeld, A. B., Tyrsted, C., Jensen, A., & Iversen, B. B. (2015). Optimized carbonation of magnesium silicate mineral for CO_2 storage. *ACS Applied Materials and Interfaces, 7*(9), 5258–5264. https://doi.org/10.1021/am508432w
46. Gerdemann, S. J., O'Connor, W. K., Dahlin, D. C., Penner, L. R., & Rush, H. (2007). Ex situ aqueous mineral carbonation. *Environmental Science and Technology, 41*(7), 2587–2593.
47. Molahid, V. L. M., Kusin, F. M., Hasan, S. N. M., Ramli, N. A. A., & Abdullah, A. M. (2021) CO_2 sequestration through mineral carbonation: Effect of different parameters on carbonation of Fe-rich mine waste materials. *Processes, 10*(2), 432.
48. Hasan, S. N. M. S., Kusin, F. M., Shamsuddin, J., & Yusuff, F. M. (2018). Potential of soil, sludge and sediment for mineral carbonation process in Selinsing gold mine, Malaysia. *Minerals, 8*, 257.
49. Mohd-Isha, N. S., Kusin, F. M., Kamal, N. M. A., Hasan, S. N. M. S., & Molahid, V. L. M. (2021). Geochemical and mineralogical assessment of sedimentary limestone mine waste and potential for mineral carbonation. *Environmental Geochemistry and Health, 43*(5), 2065–2080.
50. Liu, W., Teng, L., Rohani, S., Qin, Z., Zhao, B., Xu, C. C., Ren, S., Liu, Q., & Liang, B. (2021). CO_2 mineral carbonation using industrial solid wastes: A review of recent developments. *Chemical Engineering Journal, 416*, 129093.
51. Hitch, M., & Dipple, G. M. (2012). Economic feasibility and sensitivity analysis of integrating industrial-scale mineral carbonation into mining operations. *Minerals Engineering, 39*, 268.
52. Rahmani, O., Junin, R., Tyrer, M., & Mohsin, R. (2014). Mineral carbonation of red gypsum for CO_2 sequestration. *Energy & Fuels, 28*(9), 5953–5958.

Chapter 9
Environmental Sustainability Assessment of Alternative Controlled Low Strength Materials as a Fill Material

Anshumali Mishra⓪, Sarat Kumar Das⓪, and Krishna R. Reddy⓪

9.1 Introduction

The majority of the world's greenhouse gas (GHG) emissions (37%) are produced by the construction industry [1]. In construction operations, addressing growing environmental concerns requires prioritizing sustainable practices, including attention to small details, such as backfilling techniques, that are often overlooked in project planning [2]. Controlled low strength material (CLSM), which offers a potential sustainable solution which was not thoroughly explored, has emerged as an alternative backfilling approach in recent years [3–6]. CLSM, also referred to as flowable fill, is a permitted alternative to the typical aggregate back-filling process used to fill utility trenches and other excavations at construction sites. A CLSM is a self-leveling, highly flowable material that is often comprised of cement, water, and a fine aggregate substance like industrial waste. With a compressive strength of 8.3 MPa or less [7], the CLSM is a self-compacting, flowable, strong, and long-lasting cementitious material, according to the American Concrete Institute (ACI). The advantages of a CLSM over a conventional backfill include the absence of vibration compaction, decreased labour requirements, narrower trench, ease of distribution in complex sites (due to its flowability), low compressive strength, absence of settlement, flexibility, durability, and the mouldability with byproducts that would be disposed of in landfills. The use of cement and transportation of raw materials in CLSM mix design can

A. Mishra (✉) · S. K. Das
Indian Institute of Technology (Indian School of Mines), Dhanbad, Jharkhand, India
e-mail: anshumali.18dr0036@cve.iitism.ac.in

S. K. Das
e-mail: saratdas@iitism.ac.in

K. R. Reddy
University of Illinois Chicago, Chicago, IL, USA
e-mail: kreddy@uic.edu

S. K. Das et al. (eds.), *Geoenvironmental and Geotechnical Issues of Coal Mine Overburden and Mine Tailings*, Springer Transactions in Civil and Environmental Engineering, https://doi.org/10.1007/978-981-99-6294-5_9

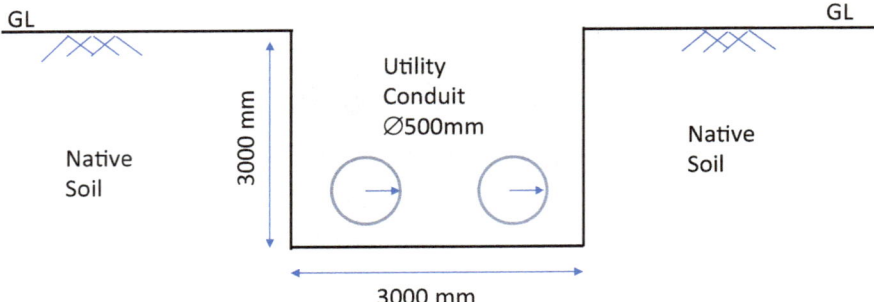

Fig. 9.1 Schematic of the proposed utility trench

increase emissions and energy consumption, which is a disadvantage of the material. This life cycle sustainability assessment's primary goals are to compare the sustainability of CLSMs to conventional backfilling methods and determine which CLSM mix design, if any, best satisfies the three pillars of sustainability (environmental, economic, and social) [8].

In this project, the environmental sustainability assessment is performed for alternative CLSMs as the backfill material in a utility trench of stretch 100 m, width 3 m and depth 3 m proposed to be constructed by Dhanbad municipal corporation as shown in Fig. 9.1. Considering, CLSM_OB [9]; ferrochrome based controlled low strength material (CLSM_FS) [10]; and wastewater treated sludge based controlled low strength material (CSLM_WTS) [11] as the alternatives adopted for the analysis are technically sound. For the baseline, conventional M5 concrete has been adopted. The different alternatives have been compared on the basis of midpoint indicators (i.e., Global warming, Ozone depletion, smog, acidification, eutrophication, carcinogenics, non-carcinogenics, respiratory effect, ecotoxicity, and fossil fuel depletion). In addition, the endpoint impacts including human health, ecosystem quality, climate change, and resource depletion have been also compared.

9.2   CSLM Properties and LCA Methodology

9.2.1   CLSM Material Properties

The material property of the alternative CLSMs is shown in Table 9.1. The design mix available in the literature are selected on the basis of strength criteria of approximately 2.5 MPa. The density adopted is the fresh density reported in the literature.

Table 9.1 Material property of alternative CLSMs

Properties	CLSM_OB [9]	CLSM_FS [10]	CLSM_WTS [11]	M5Concrete
Density (kg/cm³)	1810	2000	1742	2300
UCS (MPa)	2.5	2.65	2.44	5
CBR	135	–	–	–
Flow value	170	260	110	–
Bleeding (%)	0.25	9.73%	–	–
Mix	C:F:OB (1:2:4.5)	C:F:FS (1:0.66:7.5)	C:F:WTS (1:4:4)	C:FA:CA (1:5:10)

Note C—Cement, F—Fly ash, OB—Coalmine overburden, FS—Ferrochrome slag, WTS—Water treated sludge, FA—Fine aggregate, CA—Coarse aggregate

9.2.2 LCA Methodology

Goal and Scope

The goal of the present study is to compare different alternative backfill material on the environmental ground to find out the most sustainable alternative.

System boundary

The system boundary for the analysis includes the cradle to gate approach considering the collection of waste from the storage facility to be cradle and construction of backfill trench as the gate for the study. The end of life of the filled material is outside the scope of the present study and is not considered.

Functional unit

1 ton of CLSM is considered as the functional unit for the analysis.

Inventory Analysis

The quantity of the material used in the analysis is tabulated in Table 9.2. The quantity of the materials is estimated neglecting the volume of the conduit installed. The transport distance of cement from the source is considered as 20 km. The waste used in the study that includes OB, fly ash, ferrochrome slag, water treatment sludge and conventional aggregate were proposed to be collected from a source at a distance of 50 km. The source of water is assumed to be near the proposed site and the transportation associated with water supply is ignored.

Impact Assessment

TRACI 2.1 (Tool for Reduction and Assessment of Chemicals and Other Environmental Impacts) developed by USEPA [12] for midpoint impact categories was used for the study. Waste material emissions were not taken into account for this project;

Table 9.2 Material quantity and transport distance

Materials	Units	CSLM_OB	CLSM_FS	CLSM_WTS	M5 Concrete	Distance (km)
Cement	Tons	165	188	167	124	20
Fly ash	Tons	330	124	668	–	50
OB	Tons	742.5	–	–	–	50
Ferrochrome slag	Tons	–	1410	–	–	50
Water treatment sludge	Tons	–	–	668	–	50
Fine aggregate (FA)	Tons	–	–	–	620	50
Coarse aggregate (CA)	Tons	–	–	–	1240	50
Water	Tons	321.75	413.6	771.4	557	–

instead, emissions related to materials manufactured especially for the backfill alternative were taken into account. Impact 2002+ [13] was used for the determination of endpoint impacts in terms of human health, ecosystem quality, climate change, and resource depletion. The commercially available software SimaPro 9.4.0.2 was used for the analysis [14]. The database used for the emission inventory was Ecoinvent 3.0 [15].

9.3   Results and Discussions

9.3.1   M5 Concrete

The environmental impact caused by the use of 1 ton of M5 concrete is shown in Fig. 9.2. It is observed that the use of cement contributes relatively the most in global warming (56%), followed by smog (40%), acidification (44%), eutrophication (58%), 41% in non-carcinogenic and respiratory effect categories. The use of CA contributes relatively the most in ozone depletion (46%), carcinogenic (47%), ecotoxicity (43%), and fossil fuel depletion (45%). The production of cement involves the consumption of electricity and emission of CO_2. In contrast, the production CA involves blasting and crushing operation that may be accountable for the higher environmental impacts. The relative contribution of consumed water and cement transportation is found to be negligible. The contribution of FA and transportation of FA and CA make up the rest in each impact category. The absolute contribution of the material and transportation in the midpoint impact categories are captured in Table 9.3 and endpoint categories are captured in Table 9.4. It is observed that cement and CA contribute collectively

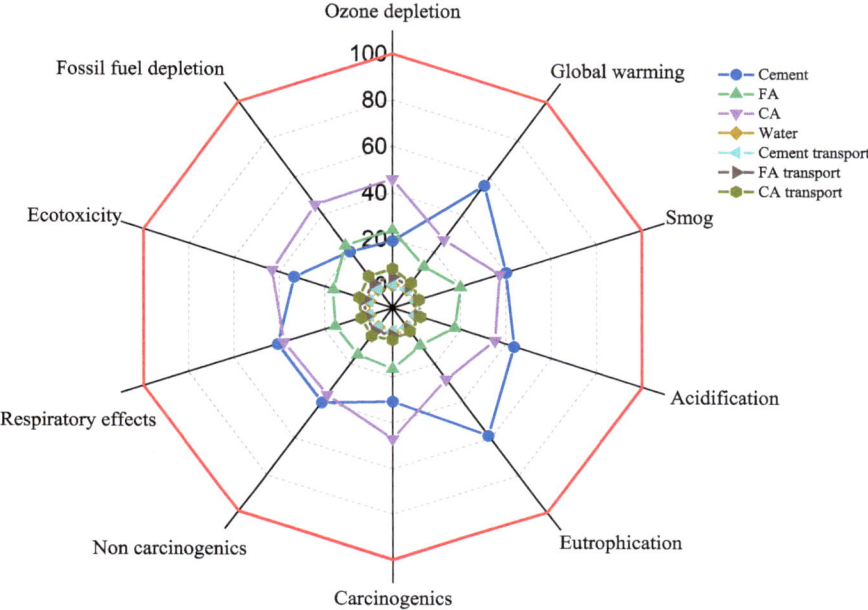

Fig. 9.2 Midpoint impact caused by the use of M5 grade of concrete as a fill material

78% impact to human health, 66% impact to ecosystem quality, 82% impact to climate change, and 73% impact on resource depletion.

9.3.2 CLSM_OB

Cement used is CLSM_OB was found to contribute relatively to most of all impact categories ranging from minimum 82% in carcinogenics and respiratory effects to maximum 95% in smog formation as shown in Fig. 9.3. The contribution of coal mine overburden (OB), water and transportation of all the materials together contribute to a minimum of 5% in smog formation and maximum of 18% in carcinogenics and respiratory effects. The absolute impact caused in the midpoint impact and endpoint impact is shown in Tables 9.5 and 9.6, respectively.

9.3.3 CLSM_FS

Similar, to CLSM_OB, the cement consumption in CLSM_FS is observed to contribute relatively the most in CLSM_FS as shown in Fig. 9.4. The absolute

Table 9.3 Absolute midpoint impacts caused due to the use of 1 ton of M5 grade concrete

Impact category	Unit	Total	Cement	FA	CA	Water	Cement transport	FA transport	CA transport
Ozone depletion	kg CFC-11 eq	1×10^{-05}	2×10^{-06}	2×10^{-06}	4×10^{-06}	1×10^{-09}	3×10^{-08}	3×10^{-07}	7×10^{-07}
Global warming	kg CO_2 eq	80	50	10	20	0.04	0.1	1	3
Smog	kg O_3 eq	7	3	1	3	0.002	0.005	0.06	0.1
Acidification	kg SO_2 eq	0.3	0.1	0.06	0.1	2×10^{-04}	3×10^{-04}	0.004	0.008
Eutrophication	kg N eq	0.1	0.07	0.01	0.03	1×10^{-04}	1×10^{-04}	0.001	0.003
Carcinogenics	CTUh	4×10^{-06}	1×10^{-06}	6×10^{-07}	2×10^{-06}	2×10^{-08}	6×10^{-09}	7×10^{-08}	1×10^{-07}
Non-carcinogenics	CTUh	1×10^{-05}	5×10^{-06}	2×10^{-06}	5×10^{-06}	1×10^{-08}	3×10^{-08}	3×10^{-07}	7×10^{-07}
Respiratory effects	kg PM2.5 eq	0.05	0.02	0.008	0.02	7×10^{-05}	8×10^{-05}	0.003	0.002
Ecotoxicity	CTUe	400	100	70	200	0.5	0.8	10	20
Fossil fuel depletion	MJ surplus	90	20	20	40	0.01	0.003	3	6

Table 9.4 Absolute endpoint impacts caused due to the use of 1 ton of M5 grade concrete

Damage category	Unit	Total	Cement	FA	CA	Water	Cement transport	FA transport	CA transport
Human health	DALY	6×10^{-05}	3×10^{-05}	1×10^{-05}	2×10^{-05}	6×10^{-08}	7×10^{-08}	9×10^{-07}	2×10^{-06}
Ecosystem quality	PDF*m^2*yr	40	10	8	10	0.009	0.1	1	3
Climate change	kg CO_2 eq	80	50	10	20	0.04	0.1	1	3
Resources	MJ primary	900	300	200	300	0.5	2	20	50

midpoint impact and endpoint impacts are shown in Tables 9.7 and 9.8, respectively. Cement contributes 95% impact in global warming and acidification, 96% impact in smog and eutrophication, 91% in respiratory effects, 90% in non-carcinogenics, 85% in carcinogenics, 88% in ecotoxicity, 76% in ozone depletion, and 77% in fossil fuel categories. The ferrochrome slag is found to contribute 21% in ozone depletion, and 20% in fossil fuels. Similarly, cement consumption is found to impact the human health (93%), ecosystem quality (80%), climate change (95%), and resources (87%).

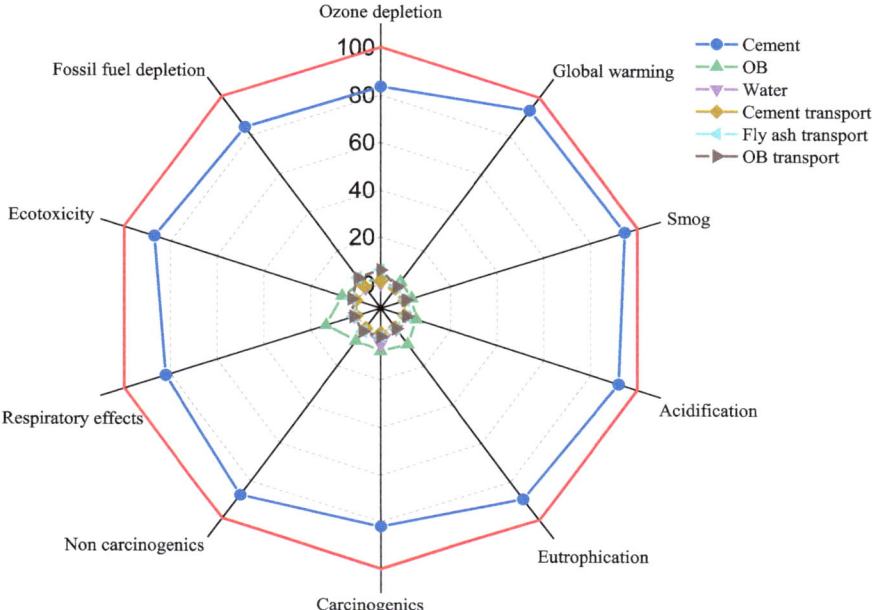

Fig. 9.3 Midpoint impact caused by the use of CLSM_OB as fill material

9.3.4 CLSM_WTS

The environmental impact caused by the use of CLSM_WTS is shown in Fig. 9.5. It is observed that the use of cement is found to contribute the maximum in all midpoint impacts, 96% in smog formation, 95% in eutrophication, 94% in global warming and acidification, 90% in respiratory effect, 89% in non-carcinogenics, 87% in ecotoxicity, 79% in carcinogenics, 77% in fossil fuel, and 76% in ozone depletion. The use of water contributes 12% in carcinogenics. The transportation of fly ash and WTS contributes 11% each in ozone depletion and fossil fuel depletion. The contribution of other waste consumption and transportation is to a maximum of 10% in other midpoint impact categories. The use of cement dominates the endpoint impacts with 92% in human health, 80% in ecosystem quality, 94% in climate change, and 87% in resource depletion. The absolute impact in endpoint categories is shown in Table 9.9, and in midpoint impacts is shown in Table 9.10.

Table 9.5 Absolute midpoint impacts caused due to the use of 1 ton of CLSM_OB

Impact category	Unit	Total	Cement	OB	Water	Cement transport	Flyash transport	OB transport
Ozone depletion	kg CFC-11 eq	4×10^{-06}	3×10^{-06}	1×10^{-07}	8×10^{-09}	5×10^{-08}	2×10^{-07}	2×10^{-07}
Global warming	kg CO_2 eq	80	80	3	0.3	0.2	0.9	0.9
Smog	kg O_3 eq	5	5	0.2	0.02	0.009	0.04	0.04
Acidification	kg SO_2 eq	0.3	0.2	0.01	0.003	5×10^{-04}	0.003	0.003
Eutrophication	kg N eq	0.1	0.1	0.01	0.001	2×10^{-04}	9×10^{-04}	9×10^{-04}
Carcinogenics	CTUh	2×10^{-06}	2×10^{-06}	2×10^{-07}	1×10^{-07}	1×10^{-08}	5×10^{-08}	5×10^{-08}
Non-carcinogenics	CTUh	1×10^{-05}	9×10^{-06}	7×10^{-07}	1×10^{-07}	4×10^{-08}	2×10^{-07}	2×10^{-07}
Respiratory effects	kg PM2.5 eq	0.04	0.04	0.006	5×10^{-04}	1×10^{-04}	7×10^{-04}	6×10^{-04}
Ecotoxicity	CTUe	300	300	20	4	1	7	7
Fossil fuel depletion	MJ surplus	40	30	1	0.1	0.4	2	2

Table 9.6 Absolute endpoint impacts caused due to the use of 1 ton of CLSM_OB

Damage category	Unit	Total	Cement	OB	Water	Cement transport	Fly ash transport	OB transport
Human health	DALY	5×10^{-05}	4×10^{-05}	5×10^{-06}	4×10^{-07}	1×10^{-07}	6×10^{-07}	6×10^{-07}
Ecosystem quality	PDF*m^2*yr	20	20	0.6	0.07	0.2	0.9	0.9
Climate change	kg CO_2 eq	80	80	3	0.3	0.2	0.9	0.9
Resources	MJ primary	600	500	40	4	3	20	20

9.3.5 Comparative Assessment of Alternative Backfill Materials

The comparative environmental sustainability of alternative backfill materials CLSM_OB, CLSM_FS, CLSM_WTS, and M5 grade concrete is shown in Fig. 9.6. It was observed that the impact of CLSM_OB in ozone depletion is 39%, smog is 71%, acidification is 80%, carcinogenics is 63%, non-carcinogenics (78%), ecotoxicity

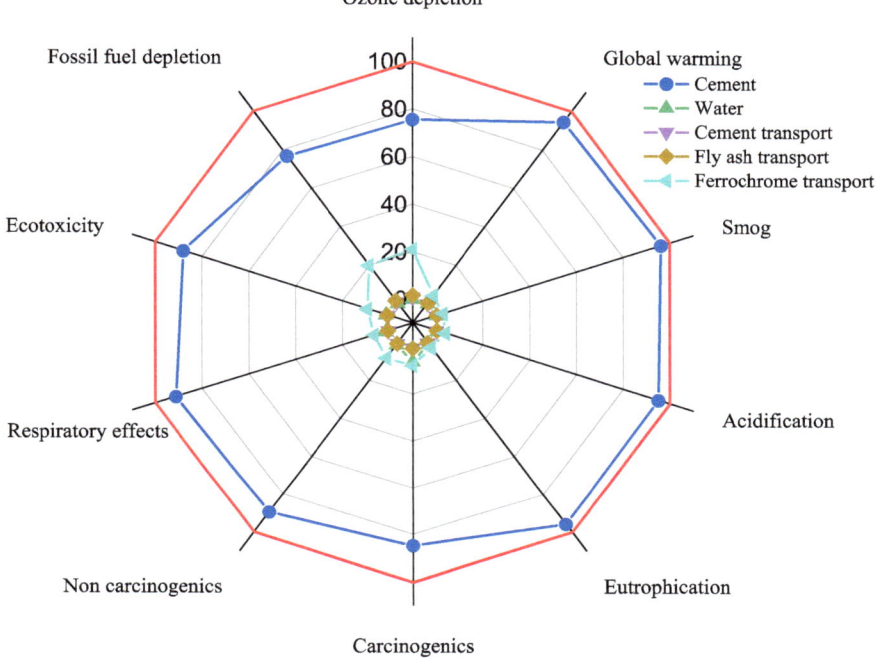

Fig. 9.4 Midpoint impact caused by the use of CLSM_FS as fill material

(65%), and fossil fuel depletion (41%). Impact in global warming (96%), respiratory effects (83%), and in eutrophication (100%) compared to other alternative filling material. In the endpoint indicators the impact of CLSM_OB is the most sustainable in ecosystem quality (52%), and resource depletion (67%). However, it is the second most in human health (80%), and climate change (96%) as shown in Fig. 9.7.

The impact caused by CLSM_ WTS is slightly more than CLSM_OB and CLSM_ FS in most of the impact categories. M5 grade concrete is found to be relatively the most polluting options among the CLSM prepared using industrial waste. The relative impact of processing OB is also shown in Figs. 9.6 and 9.7. It is highlighted that the contribution of processing is considered, however, the processing of other waste is not considered if considered that impact would have added and, in that scenario, impact caused due to processing of FS and WTS to achieve required specification for application would have added to the emissions in different impact categories.

Table 9.7 Absolute midpoint impacts caused due to the use of 1 ton of CLSM_FS

Impact category	Unit	Total	Cement	Water	Cement transport	Fly ash transport	FS transport
Ozone depletion	kg CFC-11 eq	4×10^{-06}	3×10^{-06}	9×10^{-09}	5×10^{-08}	8×10^{-08}	9×10^{-07}
Global warming	kg CO$_2$ eq	80	80	0.3	0.2	0.3	4
Smog	kg O$_3$ eq	5	5	0.02	0.009	0.01	0.2
Acidification	kg SO$_2$ eq	0.3	0.2	0.001	5×10^{-04}	9×10^{-04}	0.01
Eutrophication	kg N eq	0.1	0.1	0.001	2×10^{-04}	3×10^{-04}	0.004
Carcinogenics	CTUh	2×10^{-06}	2×10^{-06}	1×10^{-07}	1×10^{-08}	2×10^{-08}	2×10^{-07}
Non-carcinogenics	CTUh	1×10^{-05}	1×10^{-05}	1×10^{-07}	5×10^{-08}	8×10^{-08}	9×10^{-07}
Respiratory effects	kg PM2.5 eq	0.04	0.04	6×10^{-04}	1×10^{-04}	2×10^{-04}	0.003
Ecotoxicity	CTUe	300	300	4	1	2	30
Fossil fuel depletion	MJ surplus	40	30	0.1	0.4	0.7	8

Table 9.8 Absolute endpoint impacts caused due to the use of 1 ton of CLSM_FS

Damage category	Unit	Total	Cement	Water	Cement transport	Fly ash transport	FS transport
Human health	DALY	5×10^{-05}	5×10^{-05}	5×10^{-07}	1×10^{-07}	2×10^{-07}	2×10^{-06}
Ecosystem quality	PDF*m^2*yr	20	20	0.08	0.2	0.3	4
Climate change	kg CO$_2$ eq	80	80	0.3	0.2	0.3	4
Resources	MJ primary	600	500	4	3	5	60

9.4 Conclusions

The environmental sustainability of alternative backfill materials CLSM_OB, CLSM_FS, CLSM_WTS, and M5 grade concrete was assessed by TRACI 2.1 for the midpoint indicators and Impact 2002+ for the endpoint indicators. It was observed that CLSM_OB is the most sustainable in ozone depletion (39%), smog (71%), acidification (80%), carcinogenics (63%), non-carcinogenics (78%), ecotoxicity (65%), and fossil fuel depletion (41%). It is the second most sustainable in global warming

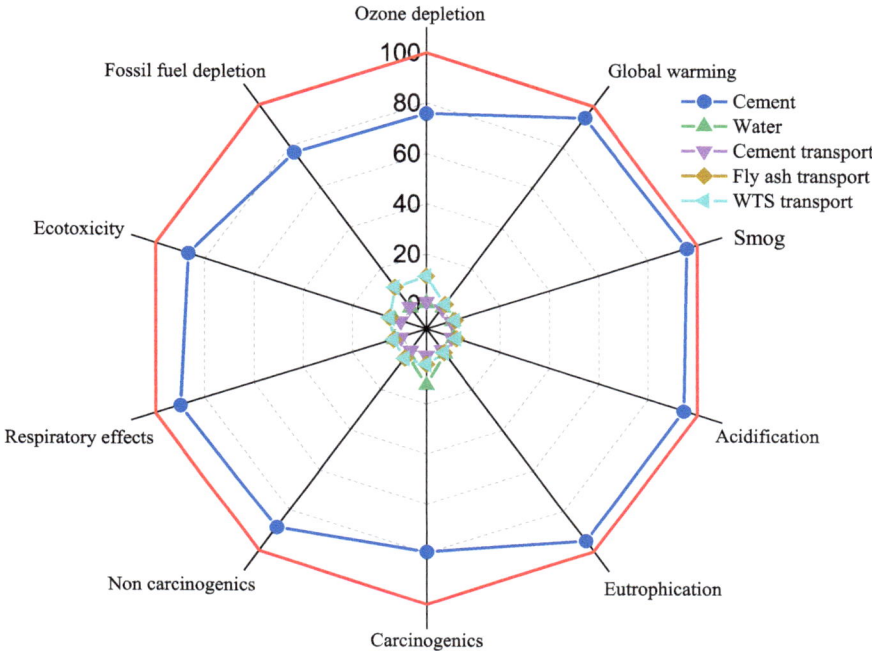

Fig. 9.5 Midpoint impact caused by the use of CLSM_WTS as fill material

Table 9.9 Absolute endpoint impacts caused due to the use of 1 ton of CLSM_WTS

Damage category	Unit	Total	Cement	Water	Cement transport	Fly ash transport	WTS transport
Human health	DALY	5×10^{-05}	5×10^{-05}	1×10^{-06}	1×10^{-07}	1×10^{-06}	1×10^{-06}
Ecosystem quality	PDF*m^2*yr	20	20	0.2	0.2	2	2
Climate change	kg CO$_2$ eq	90	80	0.7	0.2	2	2
Resources	MJ primary	600	500	9	3	30	30

(96%), respiratory effects (83%), and least sustainable in eutrophication (100%) compared to other alternative fill materials as shown in Fig. 9.6. In the endpoint indicators CLSM_OB is the most sustainable in ecosystem quality (52%), and resource depletion (67%). However, it is the second most in human health (80%), and climate change (96%).

CLSM_ FS is the second most sustainable, and CLSM_WTS is the third most sustainable. M5 grade concrete was the least sustainable compared to the CLSMs. The relative impact of processing OB is highlighted as the contribution of processing

Table 9.10 Absolute midpoint impacts caused due to the use of 1 ton of CLSM_WTS

Impact category	Unit	Total	Cement	Water	Cement transport	Fly ash transport	WTS transport
Ozone depletion	kg CFC-11 eq	4×10^{-06}	3×10^{-06}	9×10^{-09}	5×10^{-08}	8×10^{-08}	9×10^{-07}
Global warming	kg CO_2 eq	90	80	0.3	0.2	0.3	4
Smog	kg O_3 eq	6	5	0.02	0.009	0.01	0.2
Acidification	kg SO_2 eq	0.3	0.2	0.001	5×10^{-04}	9×10^{-04}	0.01
Eutrophication	kg N eq	0.1	0.1	0.001	2×10^{-04}	3×10^{-04}	0.004
Carcinogenics	CTUh	3×10^{-06}	2×10^{-06}	1×10^{-07}	1×10^{-08}	2×10^{-08}	2×10^{-07}
Non-carcinogenics	CTUh	1×10^{-05}	1×10^{-05}	1×10^{-07}	5×10^{-08}	8×10^{-08}	9×10^{-07}
Respiratory effects	kg PM2.5 eq	0.04	0.04	6×10^{-04}	1×10^{-04}	2×10^{-04}	3×10^{-03}
Ecotoxicity	CTUe	300	300	4	1	2	30
Fossil fuel depletion	MJ surplus	40	30	0.1	0.4	0.7	8

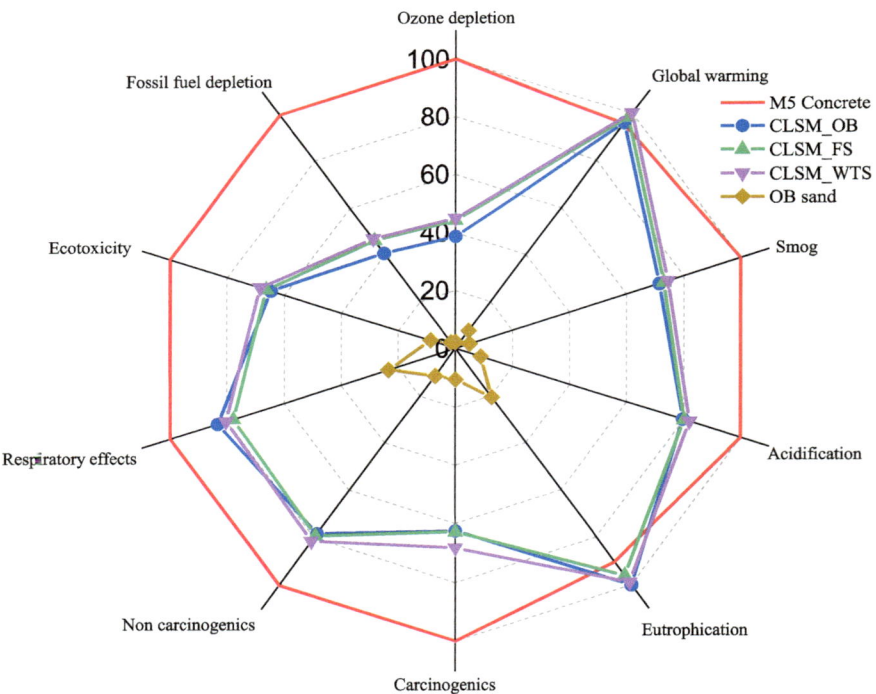

Fig. 9.6 Midpoint impact caused by the use of alternative backfill materials

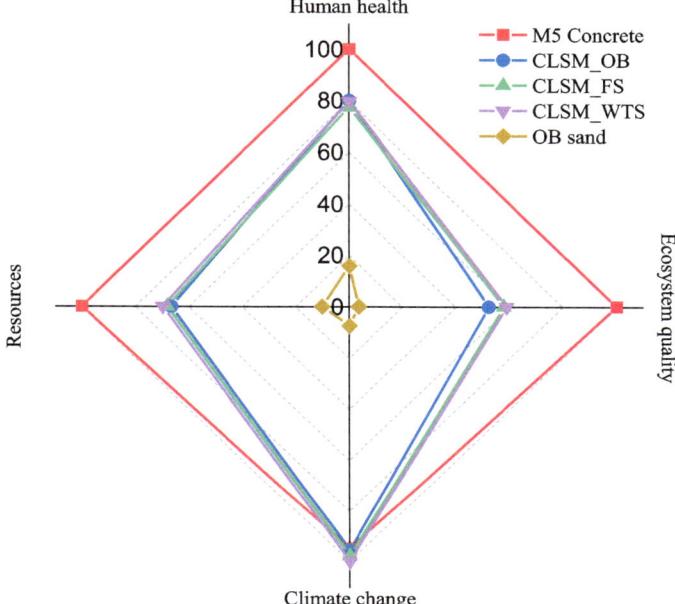

Fig. 9.7 Endpoint impact caused by the use of alternative backfill materials

OB is considered, however, the processing of other waste (FS and WTS) is not considered. If considered then the impact would have been added and, in that case, CLSM_OB would have become the most sustainable in all aspects.

References

1. UNEP. (2021). *2021 Global status report for buildings and construction: Towards a zero-emission, efficient and resilient buildings and construction sector* (pp. 1–105).
2. Alarcon, B., Aguado, A., Manga, R., & Josa, A. (2010). A value function for assessing sustainability: Application to industrial buildings. *Sustainability, 3,* 35–50.
3. Bouzalakos, S., Dudeney, A. W. L., & Chan, B. K. C. (2013). Formulating and optimising the compressive strength of controlled low-strength materials containing mine tailings by mixture design and response surface methods. *Minerals Engineering, 53,* 48–56.
4. Etxeberria, M., Ainchil, J., Pérez, M. E., & González, A. (2013). Use of recycled fine aggregates for control low strength materials (CLSMs) production. *Construction and Building Materials, 44,* 142–148.
5. Gabr, M. A., & Bowders, J. J. (2000). Controlled low-strength material using fly ash and AMD sludge. *Journal of Hazardous Materials, 76,* 251–263.
6. Naganathan, S., Razak, H. A., & Hamid, S. N. A. (2012). Properties of controlled low-strength material made using industrial waste incineration bottom ash and quarry dust. *Materials and Design, 33,* 56–63.

7. Du, L., Folliard, K. J., & Trejo, D. (2002). Effects of constituent materials and quantities on water demand and compressive strength of controlled low-strength material. *Journal of Materials in Civil Engineering, 14,* 485–495. https://doi.org/10.1061/(asce)0899-1561(2002)14:6(485)

8. Simon, S., Zanoni, L., Young, A., Hulsebosch, P., Chetri, J. K., & Reddy, K. R. (2023). *Sustainability assessment of controlled low strength materials utilizing various industrial waste materials* (vol. 280). Springer Nature Singapore. ISBN 9789811947384.

9. Das, S. K., Mahamaya, M., & Reddy, K. R. (2020). Coal mine overburden soft shale as a controlled low strength material. *International Journal of Mining, Reclamation and Environment, 34,* 725–747. https://doi.org/10.1080/17480930.2020.1721043.

10. Mahamaya, M., & Das, S. K. (2020). Characterization of ferrochrome slag as a controlled low-strength structural fill material. *International Journal of Geotechnical Engineering, 14,* 312–321. https://doi.org/10.1080/19386362.2018.1448527

11. Ho, L. S., Jhang, B. J., Hwang, C. L., & Huynh, T. P. (2022). Development and characterization of a controlled low-strength material produced using a ternary mixture of portland cement, fly ash, and waste water treatment sludge. *Journal of Cleaner Production, 356,* 131899. https://doi.org/10.1016/j.jclepro.2022.131899

12. Bare, J. C., Norris, G. A., & Pennington, D. W. (2003). The tool for the reduction and assessment impacts. *Journal of Industrial Ecology, 6,* 49–78.

13. Huijbregts, M. A. J. (1998). Application of uncertainty and variability in LCA. Part I: A general framework for the analysis of uncertainty and variability in life cycle assessment. *International Journal of Life Cycle Assessment, 3,* 273–280. https://doi.org/10.1007/BF02979835.

14. PRé Sustainability SimaPro. Retrieved April 19, 2023, from http://www.pre-sustainability.com.

15. Ecoinvent Ecoinvent 3.0. Retrieved March 19, 2023, from https://ecoinvent.org/the-ecoinvent-association/#.

Chapter 10
State of the Art Review on the Geochemical, Microbial and Environmental Aspects of Passive Acid Mine Drainage Treatment Techniques

M. K. Kaushik[ID]

10.1 Introduction

Mining impacts the environment and the legacy of mining may persist for many years long after the site has been abandoned. Another critical problem with mining is the change of chemical characteristics of natural waters around mining activity sites due to the natural oxidation of sulfide mineral tailings exposed to water, oxygen, and microorganisms. Depending on the type of interactions, namely hydrological, chemical, or biological processes, these mine water usually have a low pH which contains high levels of metals such as iron, manganese, zinc, copper, nickel, and cobalt resulting in the Acid Mine Drainage (AMD) generation which is very common at a number of sites in the Northeastern coal field (Assam, Arunachal Pradesh) and South Central coal field (Jharkhand, Orissa, Chhattisgarh) and many other places of India (Fig. 10.1). Preliminary tests conducted by the Centre of Science and Environment (CSE) in 2019 have reported that the heavy metals (like Cd, Pb, Hg, and Cr) were present in the acidic water discharges from the coal mining areas ultimately flows in the nearby rivers, water reserves results into adverse environmental effects.

Numerous recent studies [2–9] reported that when the hazardous mixture flows towards streams and rivers as well as into the groundwater produces many environmental problems. Chabukdhara and Singh [3] observed that various metals such as Fe, Mn, Ni, Pb, and Cu were present in relatively higher concentration levels in the mine waste samples from Northeastern coal fields than in other coal mining sites of India.

M. K. Kaushik (✉)
DAV Institute of Engineering and Technology, Jalandhar, Punjab 144008, India
e-mail: m.kaushik@davietjal.org

© The Author(s), under exclusive license to Springer Nature Singapore Pte Ltd. 2024
S. K. Das et al. (eds.), *Geoenvironmental and Geotechnical Issues of Coal Mine Overburden and Mine Tailings*, Springer Transactions in Civil and Environmental Engineering, https://doi.org/10.1007/978-981-99-6294-5_10

Fig. 10.1 Acid Mine Drainage (AMD) in a stream just outside of Pittsburgh, PA and Copper in natural acid steam (images by Curtis [1])

Prasad et al. [10] recorded drastic changes in the physicochemical properties of land as a result of environmental degradation. Muhammad et al. [8] showed that the quality of mine drainage affected the undistributed forest soil by means of pH, heavy metal concentration, etc. Pyrite oxidation exposed to reactive sulfides minerals in the presence of molecular O_2 causing AMD from the open pit mines ($\approx 85\%$ of the total are from the open pit as reported by Gupta and Paul [11] releases Potentially Hazardous Elements (PHEs) creates big water and soil pollution problems.

Li et al. [6] reported that Zhijin coal-mining district, located in Midwestern Guizhou Province, China, has been significantly exploited for several years. The discharge of AMD from coal mining in this area has created a severe problem for the local water environment, which greatly affected the normal use by the local people. It was also observed that the availability of macroinvertebrates in the AMD-affected areas is very low compared to unaffected areas. Also, the population of scrapers was found to be low at AMD-impacted sites due to their low tolerance capacity for contamination and due to the scarcity of benthic algae which is an important food reservoir for scrapers in the AMD-affected water [12, 13]. AMD imparts toxic effects on aquatic organisms by destroying the ecosystem and staining water in regions near the mine [14, 15]. AMD containing heavy metals has a number of serious health implications for human beings as well as for other animals due to their critical and long-term toxicity [16]. Also, it creates adverse effects on plant growth due to its low concentration of organic matter and unfavorable and poor soil structure [17]. Abandoned mines overburden also creates an adverse condition for the microbes present in the soil.

The oxidation of sulfide minerals (such as pyrite, pyrrhotite, etc.) increases the formation of sulfuric acid which consequently promotes the release of a number of traces and heavy metals like Fe, Cu, As, Cd, Hg, Co, Ni, Zn, Pb, Se, Mn, etc. [2].

10.1.1 Iron–Sulphide Oxidation

Acid mine drainage is commonly associated with the extraction and processing of sulfide-bearing metalliferous ore deposits, sulfide-rich coal, and weathering of metalliferous black shales [18]. During weathering the mobility of elements depends upon the susceptibility of minerals to chemical breakdown and subsequent release of cations and anions. The principal geo-chemical process that operates when water comes in contact with the geological mass includes oxidation–reduction, dissolution—precipitation, and ion exchange processes, etc. The oxidation of iron sulfide minerals such as pyrite ($FeS2$), pyrrhotite ($Fe1$-x $SO.7 < x < 1.0$), and chalcopyrite ($CuFeS2$) is mainly responsible for the bulk of acid production from the mining sites and mine overburdens, etc.

Pyrite Oxidation In most mining sites, Pyrite is the most widespread sulfide mineral, and weathering of pyrite is estimated at $36 * 1012$ g/year, involving 0.02 mol of electrons per square meter of earth's land-surface area. In most mine wastes, oxygen is the primary oxidizing agent of ferrous iron to ferric, which implies that sulfide oxidation generally occurs only in areas where dissolved or gaseous oxygen is present [19]. Aqueous pyrite oxidation by O2 is represented by the following overall reactions:

$$FeS_2 + 3.5O_2 + H_2O \rightarrow Fe^{2+} + 2SO_4^{2-} + 2H^+ \tag{10.1}$$

Ferrous iron (Fe^{2+}) is released into solution where it is oxidized by O_2, to ferric iron (Fe^{3+})

$$4Fe^{2+} + O_2 + 4H \rightarrow 4Fe^{3+} + 2H_2O \tag{10.2}$$

Thermodynamic considerations suggest that under anoxic conditions, ferric iron Fe^{3+}, manganese dioxide $MnO2$, and nitrate NO_3- might oxidize pyrite. There is also an opinion that in abiogenic anoxic environments, Fe^{3+} is the only relevant oxidant of pyrite. Ferric iron (Fe^{3+}) is a powerful oxidant of pyrite in highly acidic conditions [20], and reacts with pyrite according to the following reaction:

$$FeS_2 + 14Fe^{3+} + 8H_2O \rightarrow 15Fe^+ + 2SO_4^{2-} + 16H^+ \tag{10.3}$$

This reaction results in the release of 16 mol of H^+ for each mole of pyrite oxidized.

Pyrrhotite Oxidation In many mining environments, pyrrhotite is another important iron mineral often found in Cu–Ni deposits as a gangue mineral, associated with valuable ore minerals. Massive sulfide deposits usually found all around the world, more specifically pyrrhotite containing sulfide deposits are mostly available in Russia, China, Australia and Canada. Steger and Desjardins [21] reported the predominant oxidation products of pyrrhotite to be elemental sulfur and goethite as well as various sulpho-oxyanions and smaller amounts of ferric sulfate. The exposure of pyrrhotite to air leads to the consecutive formation of iron(II) oxide, an iron(III) hydroxy-oxide or hydrated iron(III) oxide. Because of its sulfidic nature, pyrrhotite dissolved rapidly under acidic conditions and generate Fe^{2+} and H_2S according to reactions (10.4) or (10.5) or more slowly via oxidative dissolution according to reactions (10.6) and (10.7):

$$Fe_{1-x}S + 2H^+ \rightarrow (1 - x)Fe^{2+} + H_2S \quad \text{(for } x = 0) \tag{10.4}$$

$$Surface > S^{2-} + 2H^+ \rightarrow H_2S \tag{10.5}$$

$$Surface > S^{2-} + 4H_2O \rightarrow SO_4^{2-} + 8H^+ + 8e^- \tag{10.6}$$

$$\text{Together with}: 2O_2 + 4H_2O + 8e^- \rightarrow 8OH^- \text{ or } 8Fe^{3+} + 8e^- \rightarrow 8Fe^{2+} \tag{10.7}$$

Oxidation of Pyrite and Pyrrhotite occurs when the mineral surface is exposed to an oxidant and water. The oxidation of pyrite proceeds to give: (1) $FeSO_4$ or $Fe_2(SO_4)_3$ (2) FeO or Fe_2O_3; or (3) pyrrhotite. Ivanov [22] has suggested three possible oxidation mechanisms for pyrrhotite at the ambient temperature to give: (1) $FeSO_4$ even with an insufficiency of oxygen; (2) $FeSO_4$ (OH) in an excess of oxygen; and (3) $Fe(OH)_3$ under the effect of sunlight.

Reactions involved during weathering of a few minerals of host rocks (aquifer matrix) in the presence of (1) sulfuric acid (as already discussed), (2) carbonic acid, (3) O_2, H_2O, and CO_2. These reactions release cations (Ca^{2+}, Mg^{2+}, Na^+ and K^+) and anions (HCO^{3-} and SO_4^{2-}) to water. Dissolutions of salts (e.g., NaCl, KCl) and minerals like gypsum also release ions into the water. Carbonic acid is derived mainly from the soil zone CO_2.

The CO_2 in the soil zone is produced as a consequence of the decay of organic matter and root transpiration, which in turn combines with downward percolating water (rainwater, surface waters) to form carbonic acid or bicarbonate according to the following reactions,

$$CO_2 + H_2O \rightarrow H_2CO_3(\text{carbonic acid}) H_2CO_3 \rightarrow H^+ + HCO_3^-(\text{bicarbonate}) \tag{10.8}$$

Bicarbonate is derived from the dissolution of carbonates and silicate minerals by the following carbonic acid-aided reactions,

$$CaCO_3(\text{calcite}) + 2H_2CO_3 \rightarrow Ca^{2+} + 2HCO_3^- \tag{10.9}$$

$$CaMg(CO_3)_2(\text{dolomite}) + 2H_2CO_3 \rightarrow Ca^{2+} + Mg^{2+} + 4HCO_3^- \tag{10.10}$$

$$2NaAlSi_3O_8(\text{albite}) + 2H_2CO_3 + 9H_2O \rightarrow$$
$$Al_2Si_2O_5(OH)_4(\text{kaolinite}) + 2Na^+ + 4H_4SiO_4 + 2HCO_3^- \tag{10.11}$$

Oxidation of sulfides generates sulfuric acid/sulfate and carbonic acid/bicarbonate according to the following reactions,

$$2FeS_2(\text{pyrite}) + 15O + 8H_2O + CO_2 \rightarrow 2Fe(OH)_3 + 4H_2SO_4 + H_2CO_3 \tag{10.12}$$

$$2FeS_2(\text{chalcopyrite}) + 17O + 6H_2O + CO_2 \rightarrow$$
$$2Fe(OH)_3 + CuSO_4 + 2H_2SO_4 + H_2CO_3 \tag{10.13}$$

$$2FeS_2 + 7O + 2H_2O \rightarrow 2Fe^{2+} + 4SO_4^{2-} + 4H^+ \tag{10.14}$$

$$4Fe^{2+} + O_2 + 4H^+ \rightarrow 4Fe^{3+} + 2H_2O \tag{10.15}$$

$$FeS_2 + 14Fe^{3+} + 8H_2O \rightarrow 15Fe^{2+} + 2SO_4^{2-} + 16H^+ \tag{10.16}$$

Sulfuric acid generated from the oxidation of sulfides can react with carbonates according to the following reactions leading to the generation of cations (Ca^{2+}, Mg^{2+}) and anions (HCO_3^-, SO_4^{2-}).

$$CaCO_3(\text{calcite}) + H_2SO_4 \rightarrow Ca^{2+} + SO_4^{2-} + H_2CO_3CaMg(CO_3)_2$$
$$+ 2H_2SO_4 \rightarrow Ca^{2+} + Mg^{2+} + 2SO_4^{2-} + 2HCO_3^-$$

Silicate minerals during weathering process react with carbonic acid or with various proportions of CO_2 and H_2O according to the following reactions and provide cations (Ca^{2+}, Mg^{2+}, Na^+, K^+) and anion (HCO_3^-) to groundwater [23, 24].

$$2NaAlSiO_8(\text{Albite}) + 2H_2CO_3 + 9H_2O \rightarrow$$
$$Al_2Si_2O_5(OH)_4(\text{kaolinite}) + 2Na^+ + 4H_4SiO_4 + 2HCO_3^- \tag{10.17}$$

or

$$CaMgFeAl_2SiO_3O_{12}(\text{Auguite}) + 6CO_2 + 5H_2O \rightarrow$$
$$Al_2Si_2O_5(OH)_4(\text{kaolinite}) + Ca^{2+} + Mg^{2+} + Fe^{2+} + 6HCO_3^- + SiO_2 \tag{10.18}$$

$$(NaK)AlSi_3O_8(\text{alkali} - \text{feldspar}) + 2CO_2 + 2H_2O \rightarrow$$
$$Al_2Si_2O_5(OH)_4(\text{kaolinite}) + 2Na^+ + 2K^+ + 2HCO_3^- + 4SiO_2 \tag{10.19}$$

$$CaAl_2Si_2O_8(\text{Calcic} - \text{feldspar}) + 2CO_2 + 3H_2O \rightarrow$$
$$Al_2Si_2O_5(OH)_4(\text{kaolinite}) + Ca^{2+} + SiO_2 + 2HCO^{3-} \tag{10.20}$$

The above reactions illustrating sulfuric acid–aided and carbonic acid–aided weathering of carbonates and carbonic acid-aided weathering of silicate minerals and generation of cations (Ca^{2+}, Mg^{2+}, Na^+, K^+) and anions (HCO_3^- and SO_4^{2-}) (Carbonic acid can react with both silicates and carbonates. Sulfuric acid reacts only with carbonates).

Singh et al. [25] studied values of molar ratios of (1) Ca^{2+}/HCO_3^-, (2) Ca^{2+}/SO_4^{2-}, (3) Mg^{2+}/HCO_3^- and (4) Mg^{2+}/SO_4^{2-} weathering of calcite and dolomite in the drain water collected from various mining sites and proved that the proportion of HCO_3^- and SO_4^{2-} present in mine drainage water reflects the relative abundance of the two sources of protons during chemical weathering of minerals. The relative importance of two major proton-producing reactions (viz., carbonation and sulfide oxidation) could be evaluated on the basis of the values of $HCO_3^-/(HCO_3^- + SO_4^{2-})$ equivalent ratio (C-ratio; [26]) of the water samples. C-ratio of 1.0 would signify 100% carbonic acid-aided weathering of carbonates and aluminosilicates involving pure dissolution and acid hydrolysis and by consuming protons from soil-sourced and atmospheric CO_2 (Table 10.1).

Values of **C-ratio** nearing zero indicates predominance of sulfuric acid-aided weathering of carbonates. Geological sources of dissolved solids of Ca^{2+} and Mg^{2+} in water are attributed mainly to (1) dissolution of calcite/dolomite/carbonates (mixing of calcite and dolomite in proportions varying from 0 to 100% of each mineral), (2) Ca^{2+} bearing and Mg^{2+} bearing silicate minerals and (3) Reverse cation reaction between ionic load of the water and host rock matrix [25]. If the ionic load of the alkaline earths in the water is provided exclusively by calcite/dolomite/carbonates (mixture of calcite and dolomite in proportions varying from 0 to 100% of each mineral) then the values of the Ca^{2+}/HCO_3^-, Ca^{2+}/SO_4^{2-}, Mg^{2+}/HCO_3^- and Mg^{2+}/SO^{2-} molar ratios of the hydrochemical composition of the water should be equal to the values of these molar ratios. C-ratio values > 0.5 suggested predominance of carbonic acid-aided weathering process. Conversely, a value of 0.5 suggested coupled reactions involving sulfuric acid–carbonic acid aided weathering of minerals on equal proportion (Table 10.1).

Table 10.1 Values of molar ratios of dissolved ions derived from weathering of calcite and dolomite

Mineral(s)	Weathering agent	Values of cation/anion molar ratio
Calcite	Carbonic acid (HCO_3^-)	$Ca^{2+}/HCO_3^- = 0.50$
Calcite	Sulfuric acid ($H_2SO_4^{2-}$)	$Ca^{2+}/SO_4^{2-} = 1.00$
Dolomite	Carbonic acid (HCO_3^-)	$Ca^{2+}/HCO_3^- = 0.25$ $Mg^{2+}/HCO_3^- = 0.25$
Dolomite	Sulfuric acid ($H_2SO_4^{2-}$)	$Ca^{2+}/SO_4^{2-} = 0.50$ $Mg^{2+}/SO_4^{2-} = 0.50$
Carbonates*	Carbonic acid (HCO_3^-)	$Ca^{2+}/HCO_3^- =$ from 0.25 to 0.50 $Mg^{2+}/HCO_3^- =$ from 0 to 0.25

*Carbonates: A mixture of calcite and dolomite in proportions ranging from 0 to 100% of each mineral (Reference: [25])

Table 10.1 *gives* Values of molar ratios of dissolved ions derived from weathering of calcite and dolomite in proportions ranging from 0 to 100% of each mineral. The C-ratio values of the water samples could be significantly used to suggest that carbonic acid-aided weathering of carbonates and silicates (ferromagnesian minerals, minerals of feldspar group etc.) as well as sulfuric acid-aided dissolution of carbonates played a significant role in the acquisition of dissolved solids (Ca^{2+}, Mg^{2+}, Na^+, K^+, HCO_3^- and SO_4^{2-}) by water [25].

10.1.2 Microbiological Process

Biochemical reactions under natural conditions, during the process of oxidation of sulfide minerals, result in the formation of sulfates ($FeSO_4$, $CuSO_4$ etc.) bacteria such as *Acidithiobacillus ferrooxidans* and *Acidithiobacillus thiooxidans* universally present in sulfide mineral-bearing mine tailing dumps and abandoned mines can generate acid. Microorganisms also could catalyze sulfide oxidation, thus greatly increasing the rate of pyrite dissolution [20]. From physiological experiments, it was observed that microorganisms can impact the rate of sulfide oxidation during the dissolution of pyrite. Edwards et al. [31] reported that oxidation of pyrite, arsenopyrite, marcasite and sphalerite takes place through a series of intermediate sulfur-bearing compounds where microbial utilization of sulfide and intermediate sulfur compounds can significantly affect the acidification along with the pyrite oxidation/dissolution rates.

Several researchers throughout the world studied the physicochemical properties of mines and the microbial populations therein (Table 10.2). Several isolates with high sulfur/iron oxidation and reduction properties have also been identified from mine soils which could be used for improving the soil and mine drainage quality.

Researchers concluded that despite the maximum concentration of sulfate and heavy metals a number of microorganisms are responsible for increasing AMD environments. Microbial activity increases the rate of AMD formation and may also be responsible for the bulk of AMD generated. Microorganisms living in such harsh conditions adopt special systems for tolerating their environment such as the ability to sustain internal cytoplasmic pH homeostasis in such acidic conditions.

Key microorganisms are also named extremophiles because of their ability to live in severe conditions. *Acidithiobacillus ferrooxidans* (one of the most studied acidophilic chemolithotrophs) grows best in the pH range of 1.5–3.5. Favorable geochemical conditions quickly develop with an acidic interface between the mineral surface and bacteria; which tend to lower pH to a level closer to the acidophilic optimum. These acidophilic autotrophic bacteria utilize sulfur compounds and ferrous ion as their energy source and reproduce through binary fission up to 108 cells/ml. Consequently, for the dissolution and mobilization of toxic metals such as copper, iron, zinc, cadmium, arsenic, and nickel in acidic solutions generated from abandoned mines, mine wastes, and tailing dumps, the Acidithiobacillus group of bacteria is responsible [58].

Table 10.2 Important European, American, African and Asian mines where physicochemical and microbial studies were undertaken by different researchers

Location		Mine	Soil pH	Sample used		Physico-chemical study	Microbial study type		References
				Temporal	Spatial		Culture dependent	Culture independent	
EUROPE	Iberian Belt, Spain	Cu	Acidic	No	Yes	Yes	No	No	Nieto et al. [27]
	Carnoules, France	Pb and Zn	Acidic	3 years	Yes	Yes	No	TRFLP#, Amplicon sequencing	Volant et al. [1]
	Botswana, Germany, and Sweden	Pyrite	Acidic	3 years	Yes	Yes	No	Clone library sequencing, qPCR#	Korehi et al. [28]
	Sokolov, Czech Republic	Coal	Neutral	6, 12, 21 and 45 years	No	Yes	No	PLFA#	Urbanov et al. [29]
	Sokolov, Czech Republic	Coal	Alkaline	1–44 Years	No	Yes	Total Count	RFLP	Chroňáková et al. [30]
AMERICA	California, US	Coal	Acidic	11 Month	No	Yes	No	FISH#	Edwards et al. [31]
	Arizona, US	Pb and Zn	Acidic	No	Yes	Yes	No	Clone library sequencing	Mendez et al. [32]
	Pennsylvania, US	Coal	Acidic	No	Yes	Yes	No	Amplicon sequencing, qPCR	Lee et al. [33]
	Wisconsin, US	Pb and Zn	Neutral	No	No	Yes	SRB	FISH	Labrenz and Banfield [34]
	Greens Creek, Alaska	Ag	Alkaline	No	Yes	Yes	SRB, SOB, MPN, IOB	No	Lindsay et al. [35]

(continued)

Table 10.2 (continued)

Location		Mine	Soil pH	Sample used		Physico-chemical study	Microbial study type			References
				Temporal	Spatial		Culture dependent	Culture independent		
	Sossego, Brazil	Cu	Neutral	No	Yes	Yes	No	Amplicon sequencing		Pereira et al. [36]
AFRICA	Mpumalanga Province, South Africa	Coal	Acidic	No	Yes	Yes	No	No		Fosso-Kankeu et al. [37]
	Jwaneng, Botswana	Diamond	Acidic	No	Yes	Yes	No	No		Murty and Karunakara [38]
	South Africa	Coal	Acidic	1–8 years	No	Yes	Total Count	PLFA		Claassens et al. [39]
ASIA	Kerman–Bahabad, Iran	Coal	Acidic	No	Yes	Yes	No			Shahabpour et al. [40]
	Kohistan, Pakistan	Pb–Zn	Acidic	No	Yes	Yes				Muhammad et al. [41]
	Zhuji City, China	Coal	Acidic	No	Yes	Yes	Total Count	No		Liao and Xie [42]
	Fankou, China	Pb–Zn	Acidic	No	Yes	Yes	No	T-RFLP		Huang et al. [43]
	Tongchang, Yinshan, and Yongping, China	Cu	Acidic	No	Yes	Yes	No	Microarray		Xie et al. [44]
	Southeast, China	Cu, Pb, Zn, Pyrite	Acidic	No	Yes	Yes	No	No		Kuang et al. [45]

(continued)

Table 10.2 (continued)

Location	Mine	Soil pH	Sample used		Physico-chemical study	Microbial study type		References
			Temporal	Spatial		Culture dependent	Culture independent	
Southeast China	Cu, Pb, Zn, Pyrite	Acidic	No	Yes	Yes	No	Meta-genomics, -transcriptomics	Hua et al. [46]
Guangdong Province, China	Poly-metals	Acidic	No	Yes	No	No	Meta-genomics, -transcriptomics	Chen et al. [47]
Antaiba, China	Coal	Neutral	2–30 Years	No	Yes	No	PLFA, Microarray, Amplicon sequencing	Li et al. [48]
Shuimuchong, China	Cu	Neutral	No	Yes	Yes	No	No	Liu et al. [49]
INDIA Madhya Pradesh	Coal	Acidic	No	No	Yes	No	No	Jamal et al. [50]
Assam	Coal	Acidic	No	Yes	Yes	No	No	Saikia et al. [51]
Orissa	Coal	Neutral	2–10 Year	No	Yes	No	RAPD#	Maharana and Patel [52]
Jharkhand	Coal	Neutral	No	Yes	Yes	No	No	Sinha and Sinha [53]
Jharkhand	Coal	Neutral	No	Yes	Yes	No	No	Masto et al. [54]
Jharkhand	Coal	Neutral	No	Yes	Yes	No	No	Pandey et al. [55]

References: [23, 56, 57]
#Random amplification of polymorphic DNA

Bacterial Influences on Acid Mine Drainage

A.ferrooxidans is a chemolithoautotrophic bacteria, due to the oligotrophic nature (low dissolved organic carbon concentration) of acidic environments, and their lack of illumination for phototrophy. Even when in vadose conditions, *A.ferrooxidans* can survive, if the rock retains the moisture and the mine is aerated [59–61]. Favorable geochemical conditions quickly develop with an acidic interface between the bacteria and the mineral surface, and pH was lowered to a level closer to the acidophilic optimum. The process proceeds through *A.ferrooxidans* exhibiting a quorum level for the trigger of AMD. At first colonization of metal sulfides, there is no AMD, and as the bacteria grow into microcolonies, AMD remains absent, then at a certain colony size, the population begins to produce a measurable change in water chemistry, and AMD escalates.

Oxidation of metal sulfide (by oxygen) is considered slow without colonization by acidophiles, particularly *Acidithiobacillus ferrooxidans* (synonym *Thiobacillus ferrooxidans*). These bacteria can accelerate pyritic oxidation by 106 times. A proposal for the rate at which *A.ferrooxidans* can oxidize pyrite is the ability to use ferrous iron to generate a ferric iron catalyst as per the following reaction:

$$Fe^{2+} + 1/4O_2 + H^+ \rightarrow Fe^{3+} + 1/2H_2O \qquad (10.21)$$

Under the above acidic conditions, ferric iron (Fe^{3+}) is a more potent oxidant than oxygen, resulting in faster pyrite oxidation rates. Other bacteria also implicated in AMD include *Leptospirillum ferrooxidans, Acidithiobacillus thiooxidans* and *Sulfobacillus thermosulfidooxidans.*

Microorganism Used in the Bioremediation of AMD

The search for ideal microbial isolates for the removal of heavy metals is still an ongoing process. The use of in-situ microbial efficacies will not only yields a greener technology but also be cost-effective. Microbial capabilities like bioleaching, oxidation/reduction reactions, and biosorption are generally utilized. Such isolates have a two-fold utility, viz., firstly, heavy metal resistant bacteria may be used for bio-mining of expensive metals, directly from ores or by taking back metals from effluents of any industrial process, and secondly, they may simply be used for bioremediation of metal-contaminated environments. Thus, the use of microbes in bioremediation plays a critical role particularly for diluted and widely spread contaminants.

Important pure cultures like *Gallionella, Acidithiobacillus, Acidiphilium, Desulfosporosinus,* etc., with thermophilic and acidophilic properties, have been isolated from mines which have potential use in industries and bioremediations. Researchers studied the microbial diversity of Indian mine soils where bacterial isolates like *Acidithiobacillus ferrooxidans, A. thiooxidans, Thiomonas sp.,* and *Desulfotomaculum nigrificans, Pseudomonas, Bacillus, Enterobacter, Cellulosimicrobium cellulans* and *Exiguobacterium* and *Staphylococcus* etc., were characterized and proposed to be used to remediate the polluted mine sites, drainage waters or the overburden dumpsites [59–63].

Techniques that are being researched with the use of microorganisms include the use of metal-immobilizing bacteria, biocontrol with the use of alkalinity-generating microorganisms' species, and bioleaching. Laboratory experiments were conducted by Sharma et al. [64] using a poorly graded sand that was artificially contaminated with lead (toxic heavy metal). *Algae Nostoc commune* (blue-green algae) was used individually and in consortium with *Bacillus (B.) sphaericus* and *Sporosarcina (S.) pasteurii* in sand. Up to 16 days, the uncontaminated and artificially contaminated soil specimens were inoculated with different consortia combinations and treated with two different cementation media concentrations. To analyze the engineering properties of the sand and the contaminant (lead) immobilization after biocementation, permeability, direct shear, unconfined compressive strength and leachability tests were conducted. The amount of calcite content was analyzed based on a calcimeter test. Using the Toxicity Characteristics Leaching Procedure (TCLP), the leachability of lead from biocemented sand was analyzed which showed 94–99.2% immobilization of lead. The formation of $CaCO_3$ and $PbCO_3$ was identified by microcharacterization, which showed immobilization of leachable lead. The results revealed that to increase strength, decrease permeability, and increased contaminant immobilization of sand's biotreatment with the consortium of *B. sphaericus* and *Nostoc commune* is an effective method.

Sharma et al. [65] also performed investigations on the immobilization of heavy metals, specifically Pb, Zn, Cr(VI), and bio-mediated calcite precipitation for strength improvement in the contaminated soils. Firstly, for each heavy metals, the toxicity resistance of bacteria against different concentrations (1000, 2000, 3000, 4000, and 5000 mg/l) was investigated. Sharma et al. [65] observed that Pb and Cr were less toxic to *Sporosarcina pasteurii* than Zn. For 18 days, contaminated soil was biotreated with *Sporosarcina pasteurii* and cementation solutions (a solution of urea and calcium chloride dihydrate). Biocemented sand specimens were subjected to testing of hydraulic conductivity, Unconfined Compressive Strength (UCS), Ultrasonic Pulse Velocity (UPV), calcite content, pH, Toxicity Characteristic Leaching Procedure (TCLP), X-Ray Diffraction (XRD), and Scanning Electron Microscopy (SEM). The heavy metal contaminated samples showed decrease in hydraulic conductivity and increase in UCS and UPV after biotreatment; however, the changes in engineering properties were found more moderate than clean biocemented sand. The conversion of Cr(VI) to Cr(III) followed by Cr_2O_3 precipitation in calcite lattice was observed. No Pb precipitate was identified in XRD results whereas Zn was precipitated as smithsonite ($ZnCO_3$). Pb and Cr(VI) immobilization was achieved up to 92% and 94%, respectively. In the contaminated biocemented sand, TCLP leaching showed Pb and Cr(VI) immobilized proportional to calcite precipitated amount, and higher calcite amounts yielded levels (within regulatory limits).

Mwandira et al. [66] performed studies to bioremediate lead, on the use the microbially induced calcium carbonate precipitation (MICP) technique in conjunction with the bacterium *Pararhodobacter* sp. Complete removal of 1036 mg/L of Pb^{2+} was achieved during the laboratory scale experiments. These results were further confirmed by scanning electron microscope (SEM) and X-ray diffraction (XRD) analysis, which indicated lead and co-precipitation of calcium carbonate ($CaCO_3$).

The unconfined compressive strength increased with an increase in injection interval with maximum unconfined compressive strength of 1.33 MPa for fine sand, 2.87 MPa for coarse sand and 2.80 MPa for mixed sand. To efficiently induce lead immobilisation the bacterial interval required is four times with a calcium and urea concentration of 0.5 M and bacterial concentration of 10^9 cfu/mL for *Pararhodobacter* sp. Mwandira et al. [66] concluded that for the lead bioremediation, very few low-cost in situ heavy metal treatment processes are available, therefore, bioimmobilization of lead by MICP has the potential for application as an eco-friendly and low-cost method for heavy metal remediation.

Microorganisms present in the filter bed materials also help in improving biological rehabilitation possibilities. Studying the microbial population in coal mines, especially the recently abandoned mines, without any anthropogenic interventions for bioremediations, also provides a basic understanding of microbial succession which may help design ideas to remediate AMD from contaminated sites.

10.2 Remediation Approaches

10.2.1 Active Acid Mine Drainage Remediation Approaches

Numerous column studies performed by different researchers showed that mixing with calcareous material with pyrite mine spoils or tailings may produce thiosulfate (a good reducing agent) which promotes acid mine drainage water neutralization in natural conditions. The oxidation of pyrite produces H_2SO_4 which in turn reacts with the calcite of the rocks and leads to the formation of gypsum. Meanwhile, where calcite is scarce, the acid attacks the clay and leached out alkalis and alumina to form Jarosite or natrojarosite.

Jarosite is usually formed under low pH conditions and an increase in pH leads to the formation of goethite. Studies also showed that the burial of neutralizing reagents below pyrite materials may protect water bodies more effectively than the application of lime to the surface. In the past many neutralizing reagents have been used, including lime (calcium oxide), calcium carbonate, slaked lime, sodium hydroxide, sodium carbonate, and magnesium oxide and hydroxide. These vary in effectiveness and cost; for example, sodium hydroxide is about nine times the cost of lime but is some 1.5 times as effective. When calcium-containing neutralizing reagents are used some removal of sulfate (as gypsum) is achieved. Although active chemical treatment can provide effective remediation of AMD, it has the disadvantages of problems with the disposal of the bulky sludge produced and high operating costs.

This has necessitated the development and use of alternative, low-cost, energy-efficient, and sustainable techniques for the optimal removal of toxic heavy metals from polluted waters. The application of bio-techniques coupled with engineering methods has recently been gaining attention and is being emphasized for the remediation and protection of the environment.

10.2.2 Passive Acid Mine Drainage Remediation Approaches

10.2.2.1 Construction Wetlands (CWs)

During the 1990s, around 500 'reed bed systems' or also known as 'root zone systems' were built in various regions of Europe. Since the 1990s, CWs were widely constructed and functioned to treat wastewater from different sources including agricultural farms, mine drainage, industrial sources, food processing, surface runoff, landfill leachate, sludge dewatering [67], and dairy farms [68–70].

The CW system contains natural processes of aquatic macrophytes that not only accumulate pollutants directly into their tissues but also act as catalysts for purification reactions that usually occur in the rhizosphere of the plants. In constructed wetlands, substrate interactions remove most metals from contaminated water.

Classifying Construction Wetlands (CWs)

The most accepted criterion for classifying CWs is on the basis of the type of wetland hydrology i.e., Free Water System (FWS) or surface flow (SF) and sub-surface flow (SSF) systems. Further SSF systems are categorized into horizontal SSF (HSSF) and vertical SSF (VSSF) systems, in accordance with the water flow direction. A CW system comprising combinations of different CW designs i.e., hybrid CW systems such as a combination of VSSF and HSSF CW or FWS with VSSF or HSSF systems is also emerging as a viable option. Hybrid CW combinations may prove to be helpful in maximum pollutant removal as well as the production of good-quality effluent [69] (Fig. 10.2).

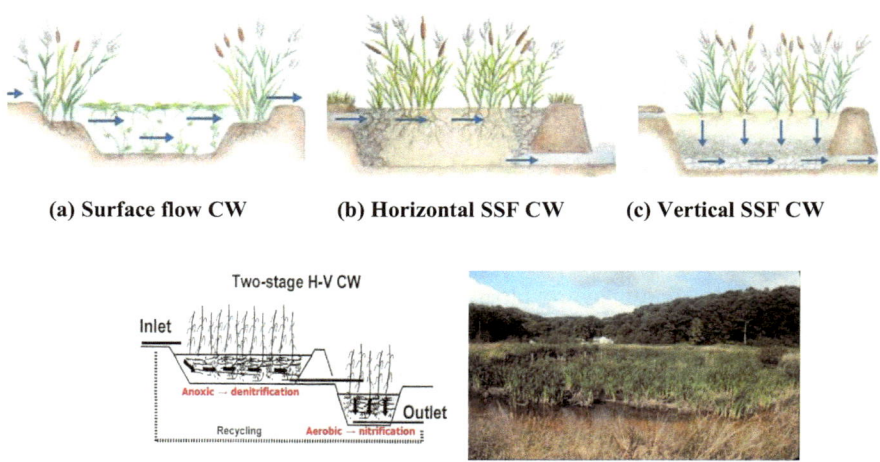

(a) Surface flow CW   (b) Horizontal SSF CW   (c) Vertical SSF CW

(d) Hybrid (Two stage) CW   (e) View of HSSF CW with Macrophytes

Fig. 10.2 Construction Wetlands (CWs) categories (based on the type of wetland hydrology)

Canna indica   *Typha Latifolia* *Phragmites australis*   *Scirpus*   *Colocasia esculenta*

Fig. 10.3 Common aquatic macrophytes communities used for Construction Wetlands (CWs)

a. Free water surface (FWS) Construction Wetlands (CWs) systems

FWS CW having emergent macrophytes plantation largely defined as a shallow closed channel or a series of channels that has generally about 20–30 cm rooting soil and 20–40 cm water depth. In the free water surface constructed wetland system settleable organic matter content is reduced by processes such as deposition and filtration. FWS CWs having emergent macrophytes plantation serve to be land-intensive biological treatment systems and the treatment methods involved are sedimentation, aggregation, and adhesion. Also, attached and suspended microbial populations play a key role in the reduction of soluble organic matter that can be decomposed aerobically as well as anaerobically. The most frequently planted macrophytes in FWS CWs include *Phragmites australis (Common reed), Typha spp. (Cattail), Scirpus spp. (Bulrush), Arundo donax (Giant reed), Sagittaria latifolia*, etc. (Fig. 10.3).

FWS CWs efficiently reduce nitrogen through nitrification/denitrification processes. The aerated zones in FWS CWs appear particularly close to the water surface due to atmospheric diffusion, and anoxic or anaerobic regions surrounding the sediments. However, this is despite numerous conflicts among civil engineers regarding odor, poor performance in cold seasons, and attraction to flies and mosquitoes.

b. Sub-Surface flow (SSF) Construction Wetlands (CWs) systems

i. Horizontal sub-surface flow Construction Wetlands (HSSF CWs)

HSSF CWs comprise of the bed filled with gravel or rock filter media substrates sealed by impervious coating and planted with wetland vegetation. In HSSF CWs, previously treated wastewater is dosed into the inlet and drifts gradually throughout the porous filter materials below the CW bed surface via a slight horizontal passageway till the water reaches the outlet for sample collection outlets (Fig. 10.2). The filtration beds allow the reduction of pollutants by several processes such as microbial activities, chemical and physical processes including a combination of simultaneous aerobic, anaerobic and anoxic reactions that are just limited to the regions surrounding the roots where oxygen leakage into the substrate occurs.

A few limitations in accordance with the records of some field measurements were also observed by various researchers such as rhizospheric oxygenation in HSSF CW system is very low and hence limited nitrification rate results in less nitrogen removal. According to the studies conducted by Vymazal and Kropfelova, horizontal treatment filters are well-known to attain incomplete nitrogen reduction that takes place because of deficient oxygen flux required to carry out the nitrification procedure.

ii. Vertical sub-surface flow Construction Wetlands (VSSF CWs)

The initially designed VSSF CW system also known as infiltration or percolation tanks with a vertical flow filled with sand or soil as a filter medium and a discharge drain at the bottom basically requires the pumping of water intermittently over the wetland bed surface. Wastewater is intermittently added onto the filter media substrate having surface vegetation and gradually flows toward the bottom of the bed. Intermittent water flow (batch-type feeding) provides good nitrification potential as well as a good oxygen transfer rate. Plant species such *as Typha, Canna, giant reed (Arundo donax), Phragmites australis,* etc. (Fig. 10.3) are commonly used as surface vegetation in VSSF CWs. The treated effluent (treated sample) is collected at the base of the CW filter beds.

VSSF CWs can prove to be the most consistent and trustworthy alternative. The infiltrate that retains after the first treatment, consists of high nutrient content (organic and inorganic matter). This quality of the infiltrate could be further improvised by operating the VSSF CW system under the recycling process either in the same or another unit.

c. Hybrid Construction Wetlands (CWs)

In order to overcome such type of limitations, it was suggested by some workers to introduce HSSF systems which have proved to be very efficient for BOD and suspended solids removal whereas VSSF CW systems provide better conditions for nitrification whereas limits denitrification efficiency. Therefore, demand for hybrid CW systems is growing day by day Besides, VSSF CW systems provide better conditions for nitrification whereas limits in denitrification efficiency is observed. Therefore, demand for hybrid CW systems is growing day by day.

In the current scenario, hybrid CWs are being operated in various nations all over the world and are generally employed for ammonium reduction and other nitrogen compounds. Further, hybrid CWs are used in the purification of different kinds of wastewater from different sources for instance, slaughterhouses, landfill leachate, compost leaching, or wineries [71]. These systems most commonly consist of VSSF and HSSF systems combined together in a staged manner.

Many previous literature works have documented the use of these treatment systems for purifying wastewater from sources such as composting leachate, dairy effluent, refinery effluent, airport runoff [71–73], etc. Some examples of VSSF CWs used in several regions of the world have been revealed in Table 10.3

In the recent past, developing countries have come up with the use of CWs in treating other different kinds of wastewater also like agricultural wastewater, lake and river water [48], sludge effluent, stormwater runoff, hospital wastewater, laboratory wastewater and landfill leachate.

Table 10.3 Types of constructed wetlands used for treatment of different kinds of wastewater

Type of constructed wetland	Type of wastewater used	Country	Reference
SF CW	Storm water runoff (residential area) Dairy pasture	Australia New Zealand	Bavor et al. [74] Tanner et al. [75]
SF CW	Cooper mine drainage Acid coal mine drainage	Canada U.S.A	Sobolewski [76] Brodie et al. [77]
SF CW	Refinery wastewater	U.S.A	Gillepsie et al. [78]
SF CW	Landfill leachate	U.S.A Sweden	Martin et al. [79] Benyamine et al. [80]
SSVFCW	River water	India	Yadav et al. [81] (Removal:- Cr 98.3%, Ni 96.2%)
SSVFCW, SSHFCW	Synthetic greywater	India	82 (Removal:- Nutrient 87.4% Heavy metal ~ 94.88%)
HF CW	Tannery	Portugal	Dias et al. [82]
HF CW	Distillery and winery	India	Billore et al. [83]
HF CW	Dairy wastewater	U.S.A	Hill et al. [84]
HF CW	Agricultural runoff	China	Zhou et al. [85]
HF CW	Landfill leachate	Solvenia	Bulc [71]
HF CW	Municipal sewage	–	Kadlec et al. [86]
HF CW	Distillery and winery	India	Billore et al. [83]
VF CW	Refinery wastewater	Pakistan	Aslam et al. [72]
VF CW	Leachate compost	Germany	Lindenblatt [87]
VF CW	Municipal/Domestic	Worldwide	Cooper et al. [88], Kadlec et al. [86]
VF CW	Refinery wastewater	Pakistan	Aslam et al. [72]
VF CW	Compost leachate	Germany	Lindenblatt [87]
VF CW	Airport runoff	Canada	McGill [89]
Hybrid (HF-VF)	Sewage	Denmark	Brix et al. [90]
Hybrid (VF-HF)	Landfill leachate	Solvenia	Bulc [71]
Hybrid CW	Fertilizer wastewater	–	Maine et al. [91] (Removal: 80%)

Constructed wetlands (CWs) provide efficient passive low cast viable treatment options with a higher percentage of sulfide reduction, removal of other metals, and alleviation of extreme acidic conditions. CWs provide sufficient surface area for physicochemical as well as biological reactions that are involved in reducing the organic, inorganic nutrient, and metal contents from wastewater [86]. Due to these characteristics, researchers from all over the world have been attracted to CWs and are implementing CWs for the purification of wastewater from different sources. Other additional benefits include the use of clean and green treatment processes, water recycling and reuse, tolerance to influent load fluctuations as well as habitat provision to various wetland life forms. Aerobic wetlands are designed to encourage the precipitation of hydroxides or metal oxides by maintaining suitable residence time and aeration period, availability of energetic microbial biomass, chemistry of mine water, conc. of dissolved metals etc.

Two lab-scale hybrid wetland system's pollutant removal efficiencies for treating textile wastewater were also studied by Saeed and Sun [92]. The systems were operated under high hydraulic loading (HL) (566–5660 mm/d), organics loadings (9840–19,680 g COD/m^2 d and 2154–4307 gBOD$_5$/m^2 d) and inorganic nitrogen (254–508 g N/m^2 d). In the first stage of VF wetlands of BOD$_5$ (74–79%) and ammonia (59–66%) instantaneous removals, demonstrated the efficiency of the media for oxygen transfer to cope with the high pollutant loads. The organic carbon (C) content of sugarcane bagasse facilitated denitrification in the VF wetlands. Under predominantly anaerobic conditions second-stage HF wetlands provided efficient color removal. Overall, under high and unsteady, pollutant loadings the wetland systems showed stable removal performances.

CWs also provide better nutrients as well as metal removal efficiency depending on the factors such as CW type, configuration, Hydraulic Retention Time (HRT) as well as the loading rate. CWs have high wastewater purification capability along with the least energy consumption and minimal operational as well as maintenance costs. Constructed wetlands mainly remove metals through adsorption to organic substances and also through acid-tolerant bacterial species. Aerobic wetlands are generally constructed to treat AMD that is net alkaline as the main remediating reaction that occurs within the constructed wetlands.

The main limiting factor observed for these systems is metal precipitates. These deposits need to be removed to allow for continued wetland operation. Additionally, current design practices for AMD wetland treatment systems are dominated by empirically derived contamination removal rates. Long-term move towards establishing a relationship between a process-driven basis (used to quantify and predict the Biological Physical, and Chemical interactions those control contaminants removal rates at full scale) with performance parameters (e.g., hydraulic loading, influent conc.'s, flow rates, removal rates, treatment efficiency, etc.) needs to be established for AMD Construction Wetlands (CWs) treatment technology.

10.2.2.2   Phytoremediation

The use of different kinds of plants to remove pollutants from the environment is a growing field of research because of the advantages of their environmental friendliness, cost-effectiveness, and the possibility of harvesting the plants for the extraction of absorbed contaminants such as metals [93–95]. Macrophytes are retained because it was accepted that plants probably remove metals into their biomass and that they supplied microaerobic zones for bacteria that may assist in the removal process. For the removal of contaminants from the environment for human benefit use of the biological mechanisms of innate plants known as Phytoremediation was adopted. It has been reported that these plants can accumulate heavy metals 100,000 times greater than in the associated water. Therefore, they have been used for nutrient and heavy metal removal from a variety of sources [17, 96, 97].

Aquatic macrophytes, have shown great potential to sequester selected heavy metals and nutrients by uptake through their plant bodies and through their root systems. Some of the plant species have the potential to grow in heavy metal-contaminated soils [98, 99]. Many factors may affect the HMs' accumulation which includes species, topography, growth conditions, stage and length of contaminant exposure, and ability to bioaccumulate [100].

Bang et al. [101] experimentally demonstrated the phytoremediation capability of Miscanthus sp. Goedae-Uksae 1, a hybrid, perennial, bio-energy crop developed in South Korea for the uptake of six metals (As, Cu, Pb, Ni, Cd, and Zn). The plant showed the highest removal for As (97.7%) and minimal removal in the case of Zn (42.9%). In addition, Goedae-Uksae 1 absorbed all the metals from contaminated water except As. Alternatively, Cd, Pb, and Zn were 100% removed from contaminated water samples. The free-water flow wetlands planted with *Phragmites australis* were the most effective in metal removal. The majority of the bacterial communities were amply dominated by the Proteo-bacteria phylum and the highest diversity and richness was found in those occupying mesocosms planted with Phragmites australis.

Yadav [102] studied the comparative evaluation of Cu, Cr, Co, Ni and Zn removal from aqueous solution for three different plants species, i.e., *Canna indica L., Typha angustifolia L.,* and *Cyperus alternifolius L.* planted in vertical constructed wetlands (CWs) microcosms. Results demonstrate that the wetland bed depth has a significant, direct effect on final heavy metal removal efficiencies leading to 16–23% increase in the removal efficiency when the gravel bed depth of CWs was increased from 0.3 to 1.5 m. Considering three different plant species, Zn removal was found to be highest and Co removal was found to be the lowest. In all the cases, the contribution of below-ground biomass (roots) was more in metal removal as compared to above-ground biomass (leaves and stem). Ondo et al. [103] reported that *Amaranthus cruentus* had greater potential for bioaccumulation for copper and zinc in their roots than in leaves and shoots (<0.0001).

Abdallah [104] carried out a study to see the efficiency of two aquatic duckweed macrophytes, *Lemna gibba* and *Ceratophyllum demersum* in eliminating Pb and Cr. These plants were grown in four concentrations in a single metal solution of the two metals and were harvested at regular intervals after 2, 4, 6, 9, and 12 days in the laboratory. Plants showed good performance in removing chromium and lead from their solution and had been capable of removing up to 95% Pb and 84% Cr. Among the two species, *L. gibba* was found to be more efficient in reducing Pb and Cr

Abdallah [104] reported that the heavy metal removal efficiency of duckweed becomes enhanced when grown in polyculture with different aquatic and sub-merged plant species. The exposure concentration and duration are effective on lipid peroxidation, ion leakage, protein content, and antioxidant enzyme activities in plants grown hydroponically in arsenic solution. The highest accumulation was obtained for the highest cadmium concentration of 10.0 mg Cd/L as 11.668 mg Cd/g at pH 6.0, and 38.650 mg Cd/g at pH 5.0. The cadmium accumulation gradually increased with the initial concentration of the medium. Wolffia has been reported to be a strong Cd accumulator and has great Cd phytoremediation potential. The enormous accumulation ability was mostly due to the passive adsorption of Cd by the apoplast.

The fast accumulation and transfer of Pb from nutritive solution to plants has been reported by Sobrino et al. (2010). Aquatic macrophytes have been reported to accumulate heavy metals from natural ecosystems (Muzamdar & Das, 2014; Shah et al., 2015) as well as from the constructed wetland systems 10,000 times greater than in the surrounding water column (Huguenot et al., 2014; Weiss et al., 2006; Yilmaz & Akbulut, 2011). Zayed [105] found that under prolonged experimental conditions, duckweed proved to be a good accumulator of Cd, Se, and Cu, a moderate accumulator of Cr, and a poor accumulator of Ni and Pb. The toxicity effect of each trace element on plant growth was in the order: Cu > Se > Pb > Cd > Ni > Cr.

Researchers also suggested that the bioconcentration factor (BCF) is a viable indicator of phytoremediation efficiency plant [104, 106]. It provides an index of the ability of the plant to accumulate the metal with respect to the metal concentration in the substrate [104]. Table 10.4 lists some common aquatic macrophytes used for heavy metal removal from aquatic ecosystems.

The heavy metal content (chromium, lead, zinc, cadmium, and iron) was analyzed in all parts (root, stem, and leaves) of the plant by Lata and Bhateria [107]. Chromium concentration was found to be high in the roots of *Amaranthus viridis* followed by stem and leaves as shown in Fig. 10.4 *Solanum nigrum, Chenopodium murale, Datura inoxia, Cannabis sativa, Parthenium hysterophorous* also accumulated considerable amounts of chromium. Obtained results also revealed that concentrations of chromium, zinc, and lead were higher in *Amaranthus viridis L.* and *Parthenium hysterophorous L.* as compared to other plants (Fig. 10.4).

Saghi et al. [108] also reported the accumulation potential of Pb by *Rapistrum rugosum* and *Sinapis arvensis* in a metal-contaminated soil. Wang et al. (2015), reported that *Medicago sativa L.* (alfalfa) accumulated huge amounts of Cd, Pb, and Zn in the roots as compared to other parts of the plant. Subhashini and Swamy [109] studied the phytoremediation potential of Abutilon indicum, *Catharanthus roseus* and *Canna indica* species. Subhashini and Swamy [109] concluded that Catharanthus roseus was found to be a good accumulator of lead. The accumulation of lead was least in the roots and highest in the stem. Lead was absorbed from the soil by the roots and was further translocated to the above-ground parts i.e., stem and leaves. Roots were found to accumulate a higher amount of lead of *Amaranthus viridis*. The concentration of cadmium was also high in *Datura inoxia* but in a lesser amount in comparison with *Amaranthus viridis* (Table 10.5).

The accumulation of various heavy metals in duckweed species affects different physiological and biological activities in plants. It may include decreased shoot growth, and root growth inhibition (Gopalapillai et al., 2013). Plants evolved several effective mechanisms for tolerating high concentrations of metals in soil. High accumulation of cadmium (about 12,320–2155 mg/g at 500 mM $CdCl_2$) was reported using Lemna species. The concentration caused a gradual decrease in plant growth,

Table 10.4 Some common aquatic macrophytes used for heavy metal removal from aquatic ecosystems

Duckweed species	Biomass type	Heavy metal	Remarks	Reference
L. gibba L	Live	Cu, Pb and Zn	Sewage wastewater Cu 100%, Pb 100%, Zn 93.6% removal was observed	El-Kheir et al. (2007)
Lemna gibba L	Live	Pb	Maximum accumulation of lead achieved on third day in the 50 mg/L test, and 100% inhibitory effect on sixth day. Proved an economical alternative treatment technique	Sobrino et al. (2010)
Lemna gibba L.	Live	Al	Effective aquatic plant for municipal secondary waste water effluent treatment, Harvesting every 2 days maintains maximum efficiency	Obek and Sasmaz (2011)
Lemna minor L	Live	Ar	Accumulation of As depends on As concentration and exposure duration	Duman et al. (2010)
Lemna minor L	Live	Cd	Cd accumulation increased with initial concentration of the medium, but the opposite trend was observed for the percentages of cadmium uptake	Uysal and Taner (2010)
S. polyrhiza	Dead	Cd and Pb	Maximum adsorption capacity of Pb (II) and Cd (II) was found at optimum pH of 4.0 and6.0, contact time of 120 min, and temperature at 20 8C	Meitei et al. (2013)
Lemna perpusilla	Dead	Pb	Dosage of 4 g L 1 of plant material in a solution with an initial pH of 4.6, an initial Pb(II) concentration of 50 mg L 1 with contact time of 210 min resulted in the maximum Pb(II) removal efficiency (above 95%)	Tang et al. (2013)
Lemna minor	Live	Cr	Cr ions showed toxic effects on plants at concentration above 2.0 mg/L and system continued to remove chromium ions with low efficiency	Uysal (2013)
L. minor L	Live	Cr, Cu, Pb	Textile wastewater Cr 33%, Cu 27%, Pb 36%	Sekomo et al. [106]

increased lipid peroxidize activity, and weakened the entire antioxidative defense mechanism [110].

In some plant species coined excluders, tolerance is achieved by preventing toxic metal uptake into root cells [111]. These plants have little potential for metal extraction. One such excluder is "Merlin," a commercial variety of red fescue (Festuca rubra), used to stabilize erosion-susceptible metal-contaminated soils.

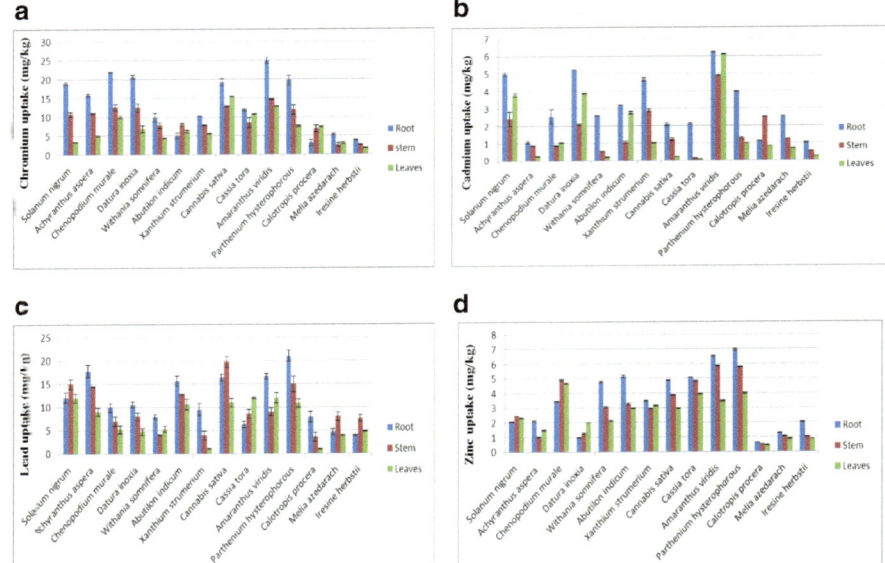

Fig. 10.4 Concentration of chromium and cadmium in various parts of the plant. Concentration of lead and Zinc in various parts of the plant [107] (*Source* Lata and Bhateria)

Table 10.5 Some common heavy metal accumulating aquatic macrophytes

Aquatic macrophytes	Common name	Heavy metal accumulation
Azolla fililiculoids	Water fern	Cr, Ni, Zn, Fe, Pb, As, Hg, Cd
Azola pinnata	Water fern	Cd, Cu, Zn, Hg
Phragmitis australis	Giant reed	Fe, Mn, Zn, Cu
Potamogeton crispus	Common reed	Cu, Pb, Mn, Fe, Cd
Salvinia spp	Water moss	Cu, Fe, Ni, Zn
Typha domingensis	Cattail	Fe, Mn, Zn, Al, Ni
Ceratophyllum demersum	Coontail	Cu, Cr, Pb, Hg, Fe, Mn, Zn, Ni
Eichhornia crassipes	Water hyacinth	Cd, Pb, Cu, As, Ni, Cr, Zn, Hg, Co, Al
Hydrilla verticillata	Water thyme	Cu, Hg, Fe, Ni, Pb
Lemna spp.	Duckweed	Pb, Mn, Cu, Cd, Cr, Hg, Ni, Fe
Mentha aquatica	Water mint	Cd, Zn, Cu, Fe, Hg
Nymphaea alba	White water lily	Cr, Cd, Pb, Ni, Zn, Mn, Fe, Co
Spirodela polyrrhiza	Giant duckweed	As, Hg

A second group of plants, coined accumulators, does not prevent metals from entering the root (Brirazani et al., 2012). Accumulator species have evolved specific mechanisms for detoxifying high metal levels accumulated in the cells. These mechanisms allow the bioaccumulation of extremely high concentrations of metals [112].

Aquatic macrophytes have been reported to accumulate heavy metals from natural ecosystems as well as from the constructed wetland systems which is about 10,000 times greater than in the surrounding water. CWs also provide better nutrients as well as metal removal efficiency depending on the factors such as CW type, configuration, loading rate as well as Hydraulic Retention Time (HRT). CWs have high wastewater purification capability along with the least energy consumption and minimal operational as well as maintenance costs.

10.3 Conclusions

Acid Mine Drainage (AMD) is commonly associated with the processing and extraction of sulfide-bearing metalliferous ore deposits, sulfide-rich coal, weathering of metalliferous black shales and sulfide oxidation of metallic/non-metallic minerals, etc. Pyrite oxidation exposed to reactive sulfide minerals in the presence of molecular O_2 causes AMD to release Potentially Hazardous Elements (PHEs). Due to its low pH (i.e., as low as pH 2) and high levels of metals (e.g., Fe, Zn, Mn, Ni, Cd, As, Pb, and Cu) and reactive sulfide content it causes harmful effects from active, abandoned mine and mine wastes on the surrounding ecosystems.

During sulfuric acid–aided and carbonic acid–aided weathering the mobility of elements depends upon the susceptibility of minerals to chemical breakdown and subsequent release of cations and anions. The proportion of HCO^{3-} and SO_4^{2-} present in mine drainage water reflects the relative abundance of the two sources of protons during the chemical weathering of minerals.

During the process of oxidation of sulfide minerals leading to the formation of sulfates ($FeSO_4$, $CuSO_4$, etc.) bacteria such as *Acidithiobacillus thiooxidans* and *Acidithiobacillus ferrooxidans* universally present in sulfide mineral-bearing mine tailing dumps and abandoned mines can generate acid producing biochemical reactions. Microorganisms living in such harsh conditions (extremophiles) adopt special systems for tolerating their environment.

Techniques that are being researched with use of microorganisms include the use of metal-immobilizing bacteria, biocontrol with the use of alkalinity-generating microorganisms' species, and bioleaching. Several isolates with high sulfur/iron oxidation and reduction properties have also been identified from mining areas soils which can be used for improving the soil and mine drainage quality by utilizing Microbial capabilities like bioleaching, oxidation/reduction reactions, and biosorption.

Constructed Wetlands (CWs) provide efficient passive low cast viable treatment options with a higher percentage of sulfide reduction, removal of other metals, and alleviation of extreme acidic conditions. Many acid-tolerant aquatic plant species such as *Typha sp.*, and Sphagnum dominate along with *Catharanthus roseus*, (*Lemna minor L., Lemna gibba L.*—duckweeds) are the most common easily available options for use in the constructed wetlands. Acid-tolerant aquatic plant species such as *Sphagnum dominates* (found to accumulate iron) along with *Ceratophyllum demersum, Lemna gibba* (accumulating Pb and Cr) and *Phragmites australis* (accumulate Fe, Mn, Zn, Cu) could effectively contribute in the remediation of acidic metal contaminated runoff waters from mines and mine wastes processing/disposal area.

A relationship between a process-driven base (used to quantify and predict the biological, physical, and chemical interactions those control contaminants removal rates) with performance parameters (e.g., hydraulic loading, influent heavy metals concentration and types, flow rates, removal rates, treatment efficiency, etc.) needs to be established for the passive AMD treatment techniques.

References

1. Volant, A. E., Bruneel, O., Desoeuvre, A., Hery, M., Casiot, C., Bru, N., Delpoux, S., Fahy, A., Javerliat, F., & Bouchez, O. (2014). Diversity and spatiotemporal dynamics of bacterial communities: Physicochemical and other drivers along an acid mine drainage. *FEMS Microbiology and Ecology, 90*, 247–263.
2. Baruah, J., Baruah, B. K., Kalita, S., & Choudhury, S. K. (2016). Physico-Chemical characteristics of drain-water of open cast coal mining area in the Ledo-Margherita range of Assam. *The Clarion, 5*(2), 30–35.
3. Chabukdhara, M., & Singh, O. P. (2016). Coal mining in northeast India: An overview of environmental issues and treatment approaches. *International Journal of Coal Science and Technology, 3*(2), 87–96.
4. Dold, B. (2014). Evolution of acid mine drainage formation in sulphidic mine tailings. *Minerals, 4*, 621–641.
5. Equeenuddin, Md. S. K., Tripathy, S., Sahoo, P. K., & Panigrahi, M. K. (2013). Metal behaviour in sediment associated with acid mine drainage stream: Role of pH. *Journal of Geochemical Exploration, 124*, 230–237.
6. L., X., Wu, P., Han, Z., Zha, X., Ye, H., & Qin, Y. (2018). Effects of mining activities on evolution of water quality of karst waters in Midwestern Guizhou, China: Evidences from hydrochemistry and isotopic composition. *Environmental Science and Pollution Research International, 25*(2), 1220–1230.
7. Saarinen, T., Mohämmadighävam, S., Marttila, H., & Klove, B. (2013). Impact of peatland forestry on runoff water quality in areas with sulphide-bearing sediments: How to prevent acid surges. *Forest Ecology and Management, 293*, 17–28.
8. Tapadar, S. A., Jha, D. K. (2015). Influence of open cast mining on the soil properties of Ledo Colliery of Tinsukia district of Assam, India. *International Journal of Scientific and Research Publications, 5*(3).
9. Vyawahre, A., & Rai, S. (2016). Acid mine drainage: A case study of an Indian Coal Mine. *IJSRSET, 2*(2)
10. Prasad, M. N. V., Nakbanpote, W., Phadermrod, C., Rose, D., Suthari, S. (2016). Chapter 13-multery and vetiver for phytostabilization of mine overburden: Overburden: Cogeneration of economic products. *Bioremediation Bioeconomy*, 295–328.
11. Gupta, A. K., & Paul, B. (2015). Ecorestoration of coal mine overburden dump to prevent environmental degradation: A review. *Research Journal of Environmental Sciences, 9*(7).

12. Jia, X. H., Jiang, W. X., Li, F. Q., Tang, T., Duan, S. G., & Cai, Q. H. (2009). The response of benthic algae to the impact of acid mine drainage. *Acta Ecologica Sinica, 29*(9), 4620–4629.
13. Niyogi, D. K., Lewis, W. M., Jr., & McKnight, D. M. (2002). Effects of stress from mine drainage on diversity, biomass, and function of primary producers in mountain streams. *Ecosystems, 5,* 554–567.
14. Ruihua, L., Lin, Z., Tao, T., & Bo, L. (2011). Phosphorus removal performance of acid mine drainage from wastewater. *Journal of Hazardous Materials, 190,* 669–676.
15. Singh, G. (1987). Mine water quality deterioration due to acid mine drainage. *International Journal of Mine Water, 6*(1), 49–61.
16. Ndlovu, S., Simate, G. S., Seepe, L., Shemi, A., Sibanda, V., & van Dyk, L. (2013). The removal of $Co2+$, $V3+$ and Cr^{3+} from waste effluents using cassava waste, South African. *Journal of Chemical Engineering, 18*(1), 1–19.
17. Rai, P. K. (2010). Seasonal monitoring of heavy metals and physicochemical characteristics in alentic ecosystem of subtropical industrial region, India. *Environmental Monitoring and Assessment, 165*(1–4), 407–433.
18. Lefticariu. (2012). www.rjes.igr.ro/wp-content/uploads/2012/07/07-sulphide-Lefticariu-24-27.pdf.
19. Benner, S. G., Gould, W. D., & Blowes, D. W. (2000). Microbial populations associated with the generation and treatment of acid mine drainage. *Chemical Geology, 169*(3–4), 435–448.
20. Nordstrom, D. K., & Alpers, C. N. (1999). Geochemistry of acid mine waters. In G. S. Plumlee & M. l. Logsdon (Eds.), *The environmental geochemistry of mineral deposits* (pp. 133–160). Society of Economic Geologists.
21. Steger, H. F., & Desjardins, L. E. (1978). Oxidation of sulphide minerals, 4. Pyrite, chalcopyrite and pyrrhotite. *Chemical Geology, 23,* 225–237.
22. Ivanov, O. P. (1966). Basic factors of development of zones of oxidation of sulfidic deposits under conditions of perennial freezing. *Geokhimiya,* 1095–1105.
23. Jones, D. S., Kohl, C., Grettenberger, C., Larson, L. N., Burgos, W. D., & Macalady, J. L. (2015). Geochemical niches of iron-oxidizing acidophiles in acidic coal mine drainage. *Applied and Environmental Microbiology, 81,* 1242–1250.
24. Dutta, M., Saikia, J., Silvio, R., Taffarel, S. R., Frans, B. W., de Medeiros, D., Cutruneo, C. M. N. L., Silva, L. F. O., & Saikia, B. K. (2017). Environmental assessment and nano-mineralogical characterization of coal, overburden and sediment from Indian coal mining acid drainage. *Geoscience Frontiers, 8,* 1285–1297.
25. Singh, A. K., Mondal, G. C., Singh, S., Singh, P. K., Singh, T. B., Tewary, B. K., & Sinha, A. (2007). Aquatic geochemistry of Dhanbad, Jharkhand: Source evaluation and quality assessment. *Journal of the Geological Society of India, 69*(5), 1088–1102.
26. Brown, G., Sharp, M., & Tranter, M. (1996). Subglacial chemical erosion: Seasonal variations in solute provenance, Haut Glacier d'Arodlla, Valais, Switzerland. *Annals of Glaciology, 22,* 25–31.
27. Nieto, J. M., Sarmiento, A. M., Canovas, C. R., Olias, M., & Ayora, C. (2013). Acid mine drainage in the Iberian Pyrite Belt: 1 Hydrochemical characteristics and pollutant load of the Tinto and Odiel rivers. *Environmental Science and Pollution Research, 20,* 7509–7519.
28. Korehi, H., Blöthe, M., & Schippers, A. (2014). Microbial diversity at the moderate acidic stage in three different sulfidic mine tailings dumps generating acid mine drainage. *Research in Microbiology, 165,* 713–718.
29. Urbanová, M., Kopecký, J., Valášková, V., Ságová-Marečková, M., Elhottová, D., Kyselková, M., Moënne-Loccoz, Y., & Baldrian, P. (2011). Development of bacterial community during spontaneous succession on spoil heaps after brown coal mining. *FEMS Microbiology and Ecology, 78,* 59–69.
30. Chronáková, A., Kristůfek, V., Tichý, M., & Elhottova, D. (2009). Biodiversity of streptomycetes isolated from a succession sequence at a post-mining site and their evidence in Miocene lacustrine sediment. *Microbiological Research, 165,* 594–608.
31. Edwards, K. J., Gihring, T. M., & Banfield, J. F. (1999). Seasonal variations in microbial populations and environmental conditions in an extreme acid mine drainage environment.

Applied and Environmental Microbiology, 65(8), 3627–3632. https://doi.org/10.1128/AEM. 65.8.3627-3632.1999.

32. Mendez, M. O., Neilson, J. W., & Maier, R. M. (2008). Characterization of a bacterial community in an abandoned semiarid lead-zinc mine tailing site. *Applied and Environmental Microbiology, 74*, 3899–3907.

33. Lee, S.-H., Sorensen, J. W., Grady, K. L., Tobin, T. C., & Shade, A. (2017). Divergent extremes but convergent recovery of bacterial and archaeal soil communities to an ongoing subterranean coal mine fire. *The ISME Journal, 11*, 1447.

34. Labrenz, M., & Banfield, J. F. (2004). Sulfate-reducing bacteria-dominated biofilms that precipitate ZnS in a subsurface circumneutral-pH mine drainage system. *Microbial Ecology, 47*, 205–217.

35. Lindsay, M. B. J., Condon, P. D., Jambor, J. L., Lear, K. G., Blowes, D. W., & Ptacek, C. J. (2009). Mineralogical, geochemical, and microbial investigation of a sulfide-rich tailings deposit characterized by neutral drainage. *Applied Geochemistry, 24*, 2212–2221.

36. Pereira, L. B., Vicentini, R., & Ottoboni, L. M. M. (2014). Changes in the bacterial community of soil from a neutral mine drainage channel. *PLoS ONE, 9*, e96605.

37. Fosso-Kankeu, E., Manyatshe, A., & Waanders, F. (2017). Mobility potential of metals in acid mine drainage occurring in the Highveld area of Mpumalanga Province in South Africa: Implication of sediments and efflorescent crusts. *International Biodeterioration & Biodegradation, 119*, 661–670.

38. Murty, V., & Karunakara, N. (2008). Natural radioactivity in the soil samples of Botswana. *Radiation Measurements, 43*, 1541–1545.

39. Claassens, S., Van Rensburg, P. J., Maboeta, M., & Van Rensburg, L. (2008). Soil microbial community function and structure in a post-mining chronosequence. *Water, Air, and Soil Pollution, 194*, 315–329.

40. Shahabpour, J., Doorandish, M., & Abbasnejad, A. (2005). Mine-drainage water from coal mines of Kerman region, Iran. *Environmental Geology, 47*, 915–925.

41. Muhammad, S., Shah, M. T., & Khan, S. (2011). Heavy metal concentrations in soil and wild plants growing around Pb–Zn sulfide terrain in the Kohistan region, northern Pakistan. *Microchemical Journal, 99*, 67–75.

42. Liao, M., & Xie, X. M. (2007). Effect of heavy metals on substrate utilization pattern, biomass, and activity of microbial communities in a reclaimed mining wasteland of red soil area. *Ecotoxicology and environmental safety, 66*, 217–223.

43. Huang, L.-N., Zhou, W.-H., Hallberg, K. B., Wan, C.-Y., Li, J., & Shu, W.-S. (2011). Spatial and temporal analysis of the microbial community in the tailings of a Pb/Zn mine generating acid drainage. *Applied and Environmental Microbiology, AEM*, 02458–02410.

44. Xie, J., He, Z., Liu, X., Liu, X., Van Nostrand, J. D., Deng, Y., Wu, L., Zhou, J., & Qiu, G. (2011). GeoChip-based analysis of the functional gene diversity and metabolic potential of microbial communities in acid mine drainage. *Applied and Environmental Microbiology, 77*, 991–999.

45. Kuang, J.-L., Huang, L.-N., Chen, L.-X., Hua, Z.-S., Li, S.-J., Hu, M., Li, J.-T., & Shu, W.-S. (2013). Contemporary environmental variation determines microbial diversity patterns in acid mine drainage. *The ISME Journal, 7*, 1038.

46. Hua, Z.-S., Han, Y.-J., Chen, L.-X., Liu, J., Hu, M., Li, S.-J., Kuang, J.-L., Chain, P. S. G., Huang, L.-N., & Shu, W.-S. (2015). Ecological roles of dominant and rare prokaryotes in acid mine drainage revealed by metagenomics and metatranscriptomics. *The ISME Journal, 9*, 1280.

47. Chen, L. X., Hu, M., Huang, L. N., Hua, Z. S., Kuang, J. L., Li, S. J., & Shu, W. S. (2015). Comparative metagenomic and metatranscriptomic analyses of microbial communities in acid mine drainage. *The ISME Journal, 9*, 1579.

48. Li, J., Liu, F., & Chen, J. (2016). The effects of various land reclamation scenarios on the succession of soil bacteria, Archaea, and Fungi over the short and long term. *Frontiers in Ecology and Evolution, 4*, 32.

49. Liu, J., Hua, Z.-S., Chen, L.-X., Kuang, J.-L., Li, S.-J., Shu, W.-S., & Huang, L.-N. (2014). Correlating microbial diversity patterns with geochemistry in an extreme and heterogeneous mine tailings environment. *Applied and Environmental Microbiology, AEM,* 00294–00214.
50. Jamal, A., Dhar, B. B., & Ratan, S. (1991). Acid mine drainage control in an opencast coal mine. *Mine Water and the Environment, 10,* 1–16.
51. Saikia, B. K., Ward, C. R., Oliveira, M. L. S., Hower, J. C., Baruah, B. P., Braga, M., & Silva, L. F. (2014). Geochemistry and nano-mineralogy of two medium-sulfur northeast Indian coals. *International Journal of Coal Geology, 121,* 26–34.
52. Maharana, J. K., & Patel, A. K. (2015). Assessment of microbial diversity associated with chronosequence coal mine overburden spoil using random amplified polymorphic DNA markers. *International Journal Recent Scientific Research, 6,* 4291–4301.
53. Sinha, M. P., & Sinha, K. (1987). *Characterisation of coal mine effluents from Jharia coalfields,* Bihar, India.
54. Masto, R. E., Ram, L. C., George, J., Selvi, V. A., Sinha, A. K., Verma, S. K., Rout, T. K., & Prabal, P. P. (2011). Impacts of opencast coal mine and mine fire on the trace elements' content of the surrounding soil vis-a-vis human health risk. *Toxicological & Environmental Chemistry, 93,* 223–237.
55. Pandey, B., Agrawal, M., & Singh, S. (2016). Ecological risk assessment of soil contamination by trace elements around coal mining area. *Journal of Soils and Sediments, 16,* 159–168.
56. Banerjee, S., Sar, A., Misra, A., Pal, S., Chakraborty, A., & Dam, B. (2018). Increased productivity in poultry birds by sub-lethal dose of antibiotics is arbitrated by selective enrichment of gut microbiota, particularly short-chain fatty acidic producers. *Microbiology, 164,* 142–153.
57. Natarajan, K. (2009). Microbial aspects of acid generation and bioremediation with relevance to Indian mining. In *Advanced Materials Research* (pp. 645–648). Trans Tech Publications.
58. Natarajan, K. A. (1998). Microbes, minerals and environment. *Geological Survey of India* (Bangalore).
59. Rehman, F., & Faisal, M. (2014). Toxic hexavalent chromium reduction by *Bacillus pumilis, Cellulosimicrobium cellulans* and *Exiguobacterium. Chinese Journal of Oceanology and Limnology, 33,* 585–589.
60. Panda, J., & Sarkar, P. (2012). Bioremediation of chromium by novel strains *Enterobacter aerogenes* T2 and *Acinetobacter* sp. PD 12 S2. *Environmental Science and Pollution Research, 19,* 1809–1817.
61. Ilias, M., Rafiqullah, I. M., Debnath, B. C., Mannan, K. S. B., & Mozammel Hoq, M. (2009). Isolation and characterization of chromium(VI)-reducing bacteria from tannery effluents. *Indian Journal of Microbiology, 51,* 76–81.
62. Desai, C., Jain, K., & Madamwar, D. (2008). Evaluation of in vitro Cr(VI) reduction potential in cytosolic extracts of three indigenous *Bacillus* sp. isolated from Cr(VI) polluted industrial landfill. *Bioresource Technology, 99,* 6059–6069.
63. Upadhyay, N., Vishwakarma, K., Singh, J., Mishra, M., Kumar, V., Rani, R., Mishra, R. K., Chauhan, D. K., Tripathi, D. K., & Sharma, S. (2017). Tolerance and reduction of chromium(VI) by *Bacillus* sp. MNU16 isolated from contaminated coal mining soil. *Frontiers in Plant Science, 8,* 778.
64. Sharma, M., Satyam, N., & Reddy, K. R. (2020). Strength enhancement and lead immobilization of sand using consortia of bacteria and Blue-Green Algae. *Journal of Hazardous, Toxic, and Radioactive Waste (ASCE), 24,* 04020049. https://doi.org/10.1061/(ASCE)HZ.2153-5515.0000548.
65. Sharma, M., Satyam, N., Reddy, K. R., & Chrysochoou, M. (2022). Multiple heavy metal immobilization and strength improvement of contaminated soil using bio-mediated calcite precipitation technique. *Environmental Science and Pollution Research, 29,* 51827–51846.
66. Mwandira, W., Nakashima, K., & Kawasaki, S. (2017). Bioremediation of lead-contaminated mine waste by Pararhodobacter sp. based on the microbially induced calcium carbonate precipitation technique and its effects on strength of coarse and fine grained sand. *Ecological Engineering, 109,* 57–64.

67. Farooqi, I. H., Basheer, F., & Chaudhari, R. J. (2008). Constructed wetland system (CWS) for wastewater treatment. In *Proceedings of Taal2007: The 12th World Lake Conference, 1004* (pp. 1009).

68. Kato, K., Inoue, T., Ietsugu, H., Koba, T., Sasaki, H., Miyaji, N., & Nagasawa, T. (2013). Performance of six multi-stage hybrid wetland systems for treating high-content wastewater in the cold climate of Hokkaido, Japan. *Ecological Engineering, 51*, 256–263.

69. Sharma, P. K., Minakshi, D., Rani, A., & Malaviya, P. (2018). Treatment efficiency of vertical flow constructed wetland systems operated under different recirculation rates. *Ecological Engineering, 120*, 474–480.

70. Sharma, P. K., Takashi, I., Kato, K., Ietsugu, H., Tomita, K., & Nagasawa, T. (2013). Effects of load fluctuations on treatment potential of a hybrid sub-surface flow constructed wetland treating milking parlor waste water. *Ecological Engineering, 57*, 216–225.

71. Bulc, T. G. (2006). Long term performance of a constructed wetland for landfill leachate treatment. *Ecological Engineering, 26*(4), 365–374.

72. Aslam, M. M., Malik, M., Baig, M. A., Qazi, I. A., & Iqbal, J. (2007). Treatment performance of compost-based and gravel-based vertical flow wetlands operated identically for refinery wastewater treatment in Pakistan. *Ecological Engineering, 30*, 34–42.

73. Vymazal, J. (2013). Emergent plants used in free water surface constructed wetlands: A review. *Ecological Engineering, 61P*, 582–592.

74. Bavor, H. J., Davies, C. M., & Sakadevan, K. (2001). Storm water treatment: Constructed wetlands yield improved pollutant management performance over a detention pond system? *Water Science and Technology., 44*(11/12), 565–570.

75. Tanner, C. C., Nguyen, M. L., & Sukias, J. P. S. (2005). Nutrient removal by a constructed wetland treating subsurface drainage from a grazed dairy pasture. *Agricultural Ecosystems and Environment, 105*, 145–162.

76. Sobolewski, A. (1996). Metal species indicate the potential of constructed wetlands for long-term treatment of mine drainage. *Journal of Ecological Engineering, 6*, 259–271.

77. Brodie, G. A., Hammer, D. A., & Tomljanovich, D. A. (1988). Constructed wetlands for acid drainage control in the Tennessee Valley. In *Proceedings Conference Mine Drainage and Surface Mine Reclamation, Vol. 1: Mine Water and Mine Waste*, U.S. Department of the Interior, Bureau of Mines, Information Circular 9183 (pp. 325–331).

78. Gillepsie, W. B., Jr., Hawkins, W. B., Rodgers, J. H., Jr., Cano, M. L., & Dorn, P. B. (2000). Transfers and transformations of zinc in constructed wetlands: Mitigation of a refinery effluent. *Ecological Engineering, 14*, 279–292.

79. Martin, C. D., Moshiri, G. A., & Miller, C. C. (1993). Mitigation of landfill leachate incorporating in-series constructed wetlands of a closed-loop design. In G. A. Moshiri (Ed.), *Constructed wetlands for water pollution improvement* (pp. 473–476). CRC Press/Lewis Publishers.

80. Benyamine, M., Bäckström, N., & Sandén, P. (2004). Multi-objective environmental management in constructed wetlands. *Environmental Monitoring and Assessment, 90*, 171–185.

81. Yadav, A., Chazarence, F., & Mutnuri, S. (2018). Development of the "French system" vertical flow constructed wetland to treat raw domestic wastewater in India. *Ecological Engineering (Elsevier), 113*, 88–93.

82. Dias, V. N., Canseiro, C., Gomes, A. R., Correia, B., & Bicho, C. (2006). Constructed wetlands for wastewater treatment in Portugal: A global overview. In *Proceedings 10th International Conference Wetland Systems for Water Pollution Control, MAOTDR 2006*, Lisbon, Portugal (pp. 91–101).

83. Billore, S. K., Singh, N., Ram, H. K., Sharma, J. K., Singh, V. P., Nelson, M. R. M., & Das, P. (2001). Treatment of a molasses based distillery effluent in a constructed wetland in central India. *Water Science and technology., 44*(11/12), 441–448.

84. Hill, C. M., Duxbury, J. M., Goehring, L. D., & Peck, T. (2003). Designing constructed wetlands to remove phosphorus from barnyard run-off: Seasonal variability in loads and treatment. In Ü. Mander & P. Jenssen (Eds.), *Constructed wetlands for wastewater treatment in cold climates* (pp. 181–196). WIT Press.

85. Zhou, Q., Zhang, R., Shi, Y., Li, Y., Paing, J., & Picot, B. (2004). Nitrogen and phosphorus removal in subsurface constructed wetland treating agriculture stormwater runoff. In *Proceedings 9th International Conference Wetland Systems for Water Pollution Control* (pp. 75–82). ASTEE 2004 and Cemagref.
86. Kadlec, R. H., Knight, R. L., Vzmayal, J., Brix, H., Cooper, P., & Haberl, R. (2000). *Constructed wetlands for pollution control—Processes, performance, design and operation, scientific and technical report* (p. 8). IWA.
87. Lindenblatt, C. (2005). Planted soil filters with activated pretreatment for composting-place wastewater treatment. In I. Toczyłowska & G. Guzowska (Eds.), *Proceedings Workshop Wastewater Treatment in Wetlands. Theoretical and Practical Aspects* (pp. 87–93). Gdańsk University of Technology Printing Office.
88. Cooper, P. F., Job, G. D., Green, M. B., & Shutes, R. B. E. (1996). *Reed beds and constructed wetlands for wastewater treatment*. WRc Publications.
89. McGill, R., Basran, D., Flindall, R., & Pries, J. (2000). Vertical-flow constructed wetland for the treatment of glycol-laden stormwater runoff at Lester B. Pearson International Airport. In *Proceedings 7th International Conference Wetland Systems for Water Pollution Control* (pp. 1080–1081). University of Florida & IWA.
90. Brix, H., Arias, C., & Johansen, N. H. (2003). Experiments in a two-stage constructed wetland system: Nitrification capacity and effects of recycling on nitrogen removal. In J. Vymazal (Ed.), *Wet lands: Nutrients, metals and mass cycling* (pp. 237–258). Backhuys Publishers.
91. Maine, M. A., Sanchez, G. C., Hadad, H. R., Caffaratti, S. E., Pedro, M. C., Mufarrege, M. M., & Di Luca, G. A. (2019). Hybrid constructed wetlands for the treatment of wastewater from a fertilizer manufacturing plant: Microcosms and field scale experiments. *Science of The Total Environment, 650*, 297–302.
92. Saeed, T., & Sun, G. (2013). A lab-scale study of constructed wetlands with sugarcane bagasse and sand media for the treatment of textile wastewater. *Bioresource Technology, 128*, 438–447.
93. Maine, M. A., Suñé, N. L., & Lagger, S. C. (2004). Chromium bioaccumulation: Comparison of the capacity of two floating aquatic macrophytes. *Water Research, 38*(6), 1494–1501.
94. Malik, A. (2007). Environmental challenge vis a vis opportunity: The case of water hyacinth. *Environment international, 33*(1), 122–138.
95. Skinner, K., Wright, N., & Porter-Goff, E. (2007). Mercury uptake and accumulation by four species of aquatic plants. *Environmental Pollution, 145*(1), 234–237.
96. Rai, P. K. (2012). An eco-sustainable green approach for heavy metals management: Two case studies of developing industrial region. *Environmental Monitoring and Assessment, 184*(1), 421–448.
97. Rawat, S. K., Singh, R. K., & Singh, R. P. (2012). Remediation of nitrite contamination in ground and surface waters using aquatic macrophytes. *Journal of Environmental Biology, 33*, 51–56.
98. Majid, N. M., Islam, M. M., & Taha, A. (2013). Heavy metal uptake and translocation in Strobilanthes crispus for phytoremediation of sewage sludge contaminated soil. *Journal of Food, Agriculture and Environment, 11*, 1514–1521.
99. Zaimoglu, Z., & Atilla, P. (2012). The uptake and translocation of hexavalent chromium and effects on growth and enzyme activity of Zea mays L. *Journal of Food, Agriculture & Environment, 10*(3&4), 982–986.
100. Ghazanfarirad, N., Dehghan, K., Fakhernia, M., Rahmanpour, F., Bolouki, M., Zeynali, F., & Bahmani, M. (2014). Determination of lead, cadmium and arsenic metals in imported rice into the west Azerbaijan province, northwest of Iran. *Journal of Novel Applied Sciences, 3*(5), 452–456.
101. Bang, J., Kamala-Kannan, S., Lee, K. J., Cho, M., Kim, C. H., Kim, Y. J., & Oh, B. T. (2015). Phytoremediation of heavy metals in contaminated water and soil using Miscanthus sp. Goedae-Uksae 1. *International Journal of Phytoremediation, 17*(6), 515–520.
102. Yadav, S. K. (2010). Heavy metals toxicity in plants: An overview on the role of glutathione and phytochelatins in heavy metal stress tolerance of plants. *South African Journal of Botany, 76*(2), 167–179.

103. Ondo, J. A., Biyogo, R. M., Abogo-Mebale, A. J., & Eba, F. (2012). Pot experiment of the uptake of metals by Amaranthus cruentus grown in artificially doped soils by copper and zinc. *Food Science and Quality Management, 9*, 28–33.
104. Abdallah, M. A. M. (2012). Phytoremediation of heavy metals from aqueous solutions by two aquatic macrophytes, Ceratophyllum demersum and Lemna gibba L. *Environmental technology, 33*(14), 1609–1614.
105. Zayed, A. (1998). Phytoaccumulation of trace elements by wetland plants duckweed. *Journal of Environmental Quality, 27*(715), 21.
106. Sekomo, C. B., Rousseau, D. P. L., Saleh, S. A., & Lens, P. N. L. (2012). Heavy metal removal in duckweed and algae ponds as a polishing step for textile wastewater treatment. *Ecological Engineering, 44*(102), 110.
107. Lata, S., & Bhateria, R. (2018). *Phytoremediation of heavy metal contaminated soil using indigenous plants.* Ph.D. Thesis, Department of Environment Sciences, Maharshi Dayanand University, Rohtak, India.
108. Saghi, A., Rashed Mohassel, M. H., Parsa, M., & Hammami, H. (2016). Phytoremediation of lead-contaminated soil by Sinapis arvensis and Rapistrum rugosum. *International journal of phytoremediation, 18*(4), 387–392.
109. Subhashini, V., & Swamy, A. V. V. S. (2015). Phytoremediation of lead, cadmium and chromium contaminated soils using selected weed plants. *Acta Biologica Indica, 4*(2), 205–212.
110. Razinger, J., Dermasti, M., Koce, J. D., Zrimec, A. (2008). Oxidative stress in duckweed (Lemna minor L.) caused by short-term cadmium exposure. *Environmental Pollution, 153,* 687–694.
111. Yang, L. T., Qi, Y. P., Jiang, H. X., & Chen, L. S. (2012). Roles of organic acid anion secretion in aluminium tolerance of higher plants. *BioMed Research International, 2013.*
112. Ahmad, A., & Al-Othman, A. A. (2014). Remediation rates and translocation of heavy metals from contaminated soil through Parthenium hysterophorus. *Chemistry and Ecology, 30*(4), 317–327.

Chapter 11
Acid Mine Drainage and Metal Leaching Potential at Makum Coalfield, Northeastern India

Sk. Md. Equeenuddin🆔, S. Tripathy🆔, Prafulla Kumar Sahoo🆔, and M. K. Panigrahi🆔

11.1 Introduction

Acid mine drainage (AMD) is a commonly occurring environmental problem associated with coal mining, and it can even last for decades after the mining has been ceased. AMD is characterized by low pH, high SO_4^{2-} and metal concentrations, particularly Fe. It is caused by the oxidation of sulfide minerals present in coal when they come in contact with atmospheric oxygen and water. AMD is one of the major sources of water pollution that affect the lotic system in numerous interactive ways resulting in serious ecological disasters [1].

Assessment of acid drainage potential is important in the management of large scale disturbances of surface or subsurface materials, especially if they contain significant amounts of sulfide minerals [2, 3]. This assessment is usually carried out through ABA study. This is the most common method for predicting the post mining water quality, and has become a widely adopted technique for overburden characterization [4, 5]. It involves the determination of acid production potential (APP) and neutralization potential (NP) of the overburden materials.

The role of carbonates and silicates in consuming acid generated during oxidation of sulfides viz. pyrite, pyrrhotite and arsenopyrite in determining the water chemistry resulted due to mining activities is well established and reviewed [6]. Carbonate

Sk. Md. Equeenuddin (✉)
National Institute of Technology, Rourkela, Odisha 769008, India
e-mail: equeen@nitrkl.ac.in

S. Tripathy · M. K. Panigrahi
Indian Institute of Technology, Kharagpur, West Bengal 721302, India

P. K. Sahoo
Central University of Punjab, Bathinda, Punjab 151401, India

minerals, mainly, calcite and dolomite are very crucial in determining the post-mining water quality and neutralize AMD and helps in inhibition of pyrite oxidation [7]. Static tests have been conducted for determining the NP of several carbonates-, aluminosilicate-minerals and with different rock types [8, 9]. The calcite and dolomite have maximum acid neutralizing capacity. Silicate minerals—pyroxene, amphibole, feldspars, micas, chlorite and clay—have relatively much lower NP, however, olivine, serpentine and wollastonite show significantly elevated NP value than the former silicate minerals. Siderite ($FeCO_3$), though it is a carbonate mineral, has zero NP value [10]. Hence, the type and occurrences of minerals at the mines help to understand the potential environmental impact due to mine discharges.

Release of metals through leaching of overburden and coal subsequent to oxidation of sulfide minerals is a potential source for contaminating water [11, 12], sediment [13] and soil [14] around both active and abandoned coal mines. Leaching of trace elements is one of major pathways for entering into the ecosystem [15]. The concentrations of various elements in both overburden and coal, and their leaching behavior are critical to understand the impact by AMD. Laboratory-scale batch leaching techniques have been widely used, and provide information on leaching potential of elements on either shorter or longer time duration in order to determine the potential impacts of mine overburden at the disposal site [16].

The occurrence of AMD at Makum coalfield of India was earlier reported and well studied [17, 18]. Beside, a significant amount of work has been carried out on the petrography [19], leaching behavior [20], metal distribution [21] and variation of sulfur in coal seams of Makum coalfield [22, 23]. However, mineralogy of overburden, ABA study of both overburden and coal, and metal leaching potential from overburden have received less attention. Therefore, an attempt has been made to study the ABA of overburden and coal from the Makum coalfield along with their metal leaching potential.

11.2 Geological Setting

Makum coalfield in Tinsukia district of Assam is the largest Tertiary coal deposit in India (Fig. 11.1) and covers about 100 km^2. It consists of three open cast coal mines, Tikak, Tirap and Ledo and two underground collieries such as Baragolai and Tipong. A total reserve of 453 million tonnes of coal has been estimated by the Geological Survey of India (unpublished, 2019).

Makum coalfield consists of five workable coal seams. Two seams having thickness of 18 m and 6 m are very prominent in the study area. Other seams are highly irregular and sporadic in nature. Coal seams belong to the Tikak Parbat Formation of Oligocene age. It comprises of sandstone, siltstone, mudstone, carbonaceous shale and coal seams. The coal is classified as semi-bituminous in rank with low ash, high sulphur (2–6%) and volatile matter content with very high caking properties [24]. Three forms of Sulphur–sulphate, pyrite and organic sulphur were observed in the Assam coal and about 70–80% of total sulphur remains in organic form [22]. Chandra

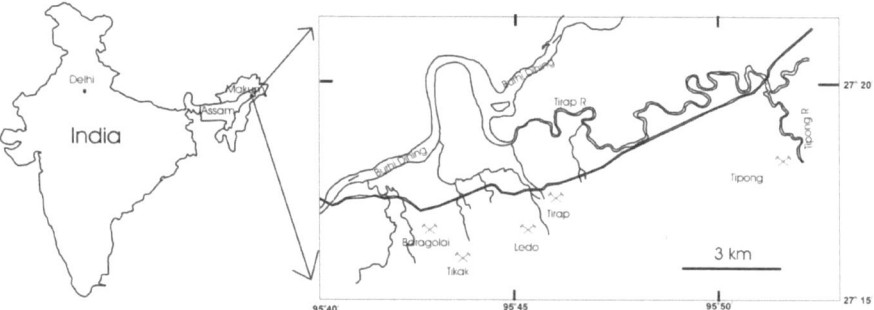

Fig. 11.1 Location of the study area and different collieries

et al. [25] and Rajarathnam et al. [23] indicated the formation of Makum coalfield under marine influence.

11.3 Materials and Methods

Seventeen overburden samples of sandstone, shale and siltstone; 13 coal samples were collected from different collieries of the Makum coalfield (Fig. 11.1). The solid samples were powdered and passed through 60 mesh for ABA test. The static ABA test was carried out by determining the NP and APP of both coal and overburden. The difference between NP and APP is termed as net neutralization potential (NNP) and the ratio of NP to APP is known as neutralization potential ratio (NPR). The NP, APP and NNP have been expressed in $CaCO_3$ equivalent tons/1000 tons (parts per thousand, ppt). The paste pH, which is a quick measure of acid generation or acid neutralization capacity of materials, was measured based on Price et al. [26] and less than 4.0 was considered potentially acid generating [27]. The paste pH was determined by placing 10 g samples in 50 ml beakers. Ultrapure water was added to the sample at 1:1 solid/solution ratio. The slurry was mixed for 5 s and pH was determined after 10 min by the pH electrode.

The NP of overburden and coal was determined following the standard Sobek method [27]. Fizz test was performed, prior to the Sobek method to determine the requisite amount and strength of HCl needed to be added for dissolving the carbonates. Fizz ratings are shown in Table 11.1. The NP was determined by adding HCl to 2 g of samples, and heated at nearly 90 to 95 °C without boiling until no bubble can be visible. After digestion, distilled water was used to bring the volume in the beaker to 100 ml. The beaker was heated for a minute and then cooled. The digested samples were titrated to pH 7 by using NaOH having normality same as that of the HCl used during the digestion step.

The SobPer method was also employed for comparison because of inability of the Sobek method to allow sufficient time for oxidation of Fe^{2+} [10]. In the SobPer

Table 11.1 Description of Fizz rating [10, 27]

Fizz rating	Description	Amount of HCl (ml)	Strength of HCl (M)
0-None	No reaction	20	0.1
1-Slight	Minimal reaction; a few bubbles per second to many fine bubbles	40	0.1
2-Moderate	Active bubbling with only a small amount of splashing	40	0.5
3-Strong	Very active bubbling that includes substantial splashing	80	0.5

method, all steps of the Sobek method are to be followed. After titration to pH 7.0, a further digestion using H_2O_2 was required for complete oxidation of Fe^{2+} as siderite is very common in the overburden and coal. 0.5 ml of 30% H_2O_2 was added to the suspension of the Sobek method and slightly boiled for 1 min. Then the solution was allowed to stand at room temperature before retitration to pH 7.0. When there was decrease in pH or the solution turned dark or green after second titration, a further H_2O_2 treatment was required. Total amount of NaOH consumed during all the titrations was used in determination of NP. However, NP measured from the SobPer method was used in all the calculations. APP was calculated by multiplying 31.25 with wt% of pyritic sulphur. This is based on the assumption that sulphide sulphur is acid generating; and sulphate and organic sulphur are nonacid generating [23].

Overburden and coal samples were sieved through 230 mesh for the detailed mineralogical study carried out by X-ray powder diffraction (XRD) using CuKα and CoKα targets. The sulphide-sulphur content of coal and overburden was measured using ASTM D 2492 and [29] methods respectively. The samples were powdered and sieved (<54 μm) for chemical analysis. Major oxides and trace elements such as Cr, Cu, Mn, Ni, Pb and Zn in overburden were determined using Philips PW 2400 Wavelength Dispersive X-ray fluorescence spectrometer (WDXRF) with Rh target using press pellets. Major oxides of coal were determined from ashed coal (850 °C) samples using WDXRF. Major elemental concentrations were determined from fused beads of ashed coal while total S analysis was done on pressed pellets of whole coal prepared with wax as the binding material. The concentrations of Cr, Cu, Ni, Mn, Pb and Zn in coal were determined by AAS (Perkin Elmer Aanalyst 300) from the digested whole coal using a mixture of HNO_3, H_2O_2 and HF.

Leaching of overburden and coal was carried out at liquid to solid ratio of 20:1 following USEPA (1994) in order to access the potential release of metals under natural weathering. The mixture of powdered overburden with deionized water (pH maintained at 4.2 by adding sulphuric and nitric acids) was gently shaken in an incubator shaker prior to measuring the pH in unfiltered splits collected after 1, 2, 4, 6 and 9 day intervals. Subsequently, the filtrates were analysed for determining the concentrations of the selected metals employing an AAS.

Fig. 11.2 XRD patterns of some overburden materials

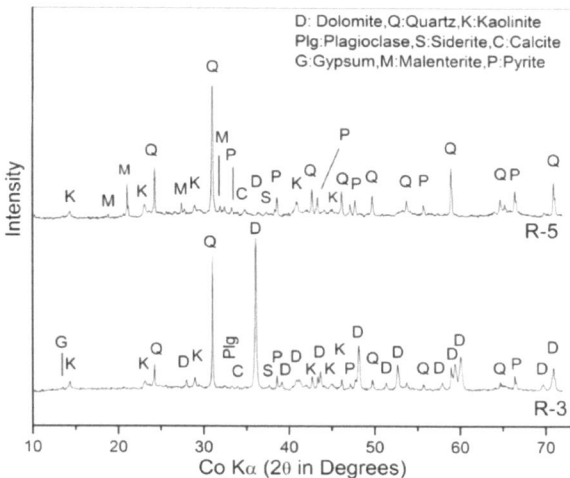

11.4 Results and Discussion

11.4.1 Mineralogy

The XRD patterns of overburden materials from the Makum coalfield are shown in Fig. 11.2. Quartz is the dominant mineral in overburden followed by kaolinite and siderite in most of the samples. However, in some overburden, dolomite is the most dominant mineral. Plagioclase has moderate occurrence. Pyrite, calcite, gypsum, melanterite and muscovite occur as minor quantity in most of the samples though in some samples pyrite was present in moderate quantity. Dolomite is abundant in overburden from the Baragolai and Ledo collieries. At Baragolai, Tirap and Ledo collieries, some overburden materials show prominent small peak at 7.6 Å which is assigned to gypsum. The overburden material at Tirap colliery shows sharp and prominent peaks of melanterite at 5.49, 4.92, 3.78, 2.27 Å. The detail mineralogy of overburden samples are given in Table 11.2. From the mineralogical study it is observed that pyrite is associated with the laminated carbonaceous shale, whereas dolomite is the dominant mineral in siltstone and non-laminated carbonaceous shale. The splintery shale, sandstones and sandy shale are found to contain trace amount of carbonates.

11.4.2 Acid Base Accounting

Various results of ABA test are given in Table 11.3. It is observed that the NP measured by the SobPer method is always lower than that of the Sobek method. A

Table 11.2 Bulk mineralogy (wt.%) of the overburden materials

Sample	Type	Quartz	Kaolinite	Dolomite	Calcite	Siderite	Pyrite	Plagioclase	Muscovite	Gypsum	Melanterite
R-1	Carb. Sh	54	20	1	2	16	–	7	–	–	–
R-2	Sst	88	9	–	–	–	–	3	–	–	–
R-3	SiS	29	10	45	–	5	8	–	1	2	–
R-4	NL Carb. Sh	31	5	60	–	–	3	–	–	1	–
R-5	L Carb. Sh	43	18	1	2	1	25	2	–	1	7
R-6	Sp. Sh	53	22	–	1	15	–	6	3	–	–
R-7	Argl. Sst	70	22	1		7	–	–	–	–	–
R-8	Sst	66	13	–	1	12	–	8	–	–	–
R-9	L Carb. Sh	63	17	–	1	3	8	1	7	–	–
R-10	Sandy Sh	83	14	1	–	2	–	–	–	–	–
R-11	Sst	55	22	–	–	17	–	6	–	–	–
R-12	SiS	29	7	52	–	9	–	3	–	–	–
R-13	Carb. Sh	74	20	–	1	3	–	–	1	1	–
R-14	Sst	84	8	–	–	1	–	7	–	–	–
R-15	Sst	83	9	–	–	3	–	4	1	–	–
R-16	Carb. Sst	65	22	–	–	12	–	2	–	–	–
R-17	Carb. Sh	77	15	5	1	–	–	2	1	–	–

Note – indicates the absence of mineral/not identified using XRD. Argl. Sst: Argillaceous Shale; Carb. Sh: Carbonaceous Shale; NL Carb. Sh: Non-laminated Carbonaceous Shale; L Carb. Sh: Laminated Carbonaceous Shale; Carb. Sst: Carbonaceous Sandstone; Sandy Sh: Sandy Shale; Sp. Sh: Splintery Shale; SiS: Siltstone; Sst: Sandstone

similar trend was noticed by several researchers [3, 10, 30], and attributed to the insufficient time for oxidation of Fe^{2+} to Fe^{3+} in the Sobek method. Since most of the samples of overburden and coal contain siderite, Sobek method has resulted in the overestimation of NP, but SobPer method is found to give more accurate NP due to complete oxidation achieved by addition of H_2O_2.

It is observed that the NP value in the overburden ranges from −67.92 to 580 ppt; APP between 0.09 and 41.25 ppt, NNP between −109.1 and 579.9 ppt and NNR between −1.65 and 6444 in the overburden. Further, very high NP of overburden is associated with Baragolai and Ledo collieries while it is intermediate at Tipong, and low at both the Tikak and Tirap. Very high NP values can be attributed to the presence of relatively higher amounts of dolomite and trace amount of calcite which is corroborated with high concentration of CaO and MgO which is up to 20.7 and 9.3% respectively. At Baragolai, it is observed that acidity generated by the oxidation of pyrite (8%) is neutralized by the presence of dolomite (48%). The negative NP at Tikak and Tirap is due to the oxidation of pyrite that is present up to 25% of the total mineral content while trace amounts of dolomite and calcite along with high amounts of quartz and kaolinite are found to be insufficient to neutralize the acid produced. The presence of pyrite is also associated with high concentration of SO_3 up to 5.8%. Melanterite which is a common efflorescent salt and often the first mineral to be deposited from aqueous solution at sites of pyrite oxidation has been observed in overburden from Tirap.

The NP value for coal varies from −162.7 to 8.62 ppt. The lowest NP is associated with Tipong and the highest is found at Ledo colliery. The APP ranges between 10.93 and 45.31 ppt with maximum at Tipong. The NNP is negative for all the coal and ranges from −3.57 to −208 ppt while NPR varies from −3.59 to 0.70. The relatively higher APP is possibly due to the abundance of reactive framboidal and very finely disseminated pyrite [17, 23]. All the coal samples contain more than 0.3 wt.% of pyritic-sulphur which is considered to be the threshold for generating AMD [26]. The pyritic-sulphur concentration in coal is above the threshold and ranges from 0.35 to 1.45 wt.% where as in overburden it lies between 0.003 and 1.32 wt.%. Therefore, the post-mining discharge quality is largely depends on the concentration of pyrite in both coal and overburden, and their neutralization potential.

A strong positive correlation between APP and S is observed which indicates that the APP is related to the pyrite content and high concentrations of S (Fig. 11.3a). On the other hand, high NP is associated with large amounts of dolomite besides traces of calcite in overburden. It is further established that high NP is due to high concentrations of Ca and Mg as strong positive correlation between Ca and NP (Fig. 11.3b), and Ca and Mg have been observed (Fig. 11.3d). No strong correlation between AP and Fe is found (Fig. 11.3c). It is possibly due to the fact that Fe is not only present as pyrite but also as siderite, which does not contribute to AP.

It has been reported that the pH of mine discharges at Baragolai and Ledo collieries are found to be mildly alkaline (Table 11.3). It is due to abundance of dolomite and calcite. Mine discharges from the Tikak and Tirap coal mines are highly acidic which is attributed to the more pyrite content in overburden. The pH of the mine discharge from Tipong ranges from 2.3 to 4.2 [17].

Table 11.3 Various parameters of acid base accounting test [12]

Colliery	Sample	Type	Paste pH	^{a}NP (npt)	^{b}NP (μμt)	Pyrite-Sulfur	APP (ppt)	NNP (ppt)	NPR	pH of mine discharge^c
Baragolai	R-1	Carbonaceous shale	8.8	54.95	38.73	0.051	1.59	37.14	24.35	7.4
	R-2	Sandstone	6.9	3.96	3.12	0.007	0.21	2.91	14.85	
	R-3	Carbonaceous shale	8.1	547.9	482	0.894	27.93	454.07	17.25	
	R-4	Carbonaceous shale	9.2	693.3	658	0.33	10.31	647.6	63.82	
	Coal		1.8	−2.27	−4.15	0.53	16.56	**−20.71**	**−0.25**	
	Coal		2.7	4.55	3.7	0.47	14.69	**−10.99**	**0.25**	
	Coal		2.2	−0.45	−0.87	0.35	10.93	**−11.8**	**−0.08**	
Tirap	R-5	Carbonaceous shale	1.9	−51.2	−67.92	1.32	41.25	**−109.1**	**−1.65**	2.5
	R-6	Splintery shale	8.7	17.8	11.86	0.035	1.09	10.77	10.88	
	R-7	Argillaceous sandstone	8.4	15.45	10.59	0.087	2.72	7.87	3.89	
	Coal		1.8 ara>	−69.4	−85.32	0.78	24.37	**−109.7**	**−3.5**	
	Coal		2.4	−2.27	−3.47	1.23	38.44	**−41.91**	**−0.09**	
Tikak	R-8	Sandstone	8.5	16.04	9.41	0.012	0.375	9.03	25.09	2.4
	R-9	Carbonaceous shale	2.2	−2.73	−3.65	0.78	24.38	**−28.03**	**−0.15**	
	R-10	Sandy shale	8.6	24.3	17.18	0.017	0.53	16.65	32.41	
	R-11	Sandstone	7.9	12.5	4.3	0.041	1.28	3.02	3.35	
	Coal		2.1	−7.5	−10.32	0.63	19.68	**−30.0**	**−0.52**	

(continued)

Table 11.3 (continued)

Colliery	Sample	Type	Paste pH	aNP (ppt)	bNP (ppt)	Pyrite-Sulfur	APP (ppt)	NNP (ppt)	NPR	pH of mine discharge[c]
Ledo	Coal		1.8	−10.2	−12.8	0.74	23.13	**−35.93**	**−0.55**	
	Coal		1.9	−14.1	−17.24	1.14	35.65	**−52.87**	**−0.48**	7.6
	R-12	Siltstone	9.6	662.7	580	0.003	0.09	579.9	6444	
	R-13	Carbonaceous shale	6.4	20.66	12.75	0.126	3.94	8.81	3.23	
	R-14	Sandstone	7.4	10.3	7.24	0.016	0.50	6.74	14.48	
	Coal		2.2	−1.14	−2.1	0.73	22.81	**−24.91**	**−0.09**	
	Coal		1.7	0.91	0.6	0.46	14.37	**−13.77**	**0.04**	
	Coal		2.8	11.13	8.62	0.39	12.19	**−3.57**	**0.70**	
Tipong	R-15	Sandstone	7.3	5.49	3.2	0.026	0.82	2.38	3.9	2.3–4.2
	R-16	Carbonaceous sandstone	7.1	11.87	6.43	0.094	2.94	3.49	2.18	
	R-17	Carbonaceous shale	8.8	76.04	62.25	0.032	1.0	61.25	62.25	
	Coal		1.4	−143.8	−162.7	1.45	45.31	**−208.0**	**−3.59**	
	Coal		1.9	−29.5	−35.2	0.85	26.56	**−61.76**	**−1.32**	

aNP determined by Sobek Method; bNP determined by SobPer Method; Bold letters indicate materials which are acid generating; [c]Equeenuddin et al. [17]

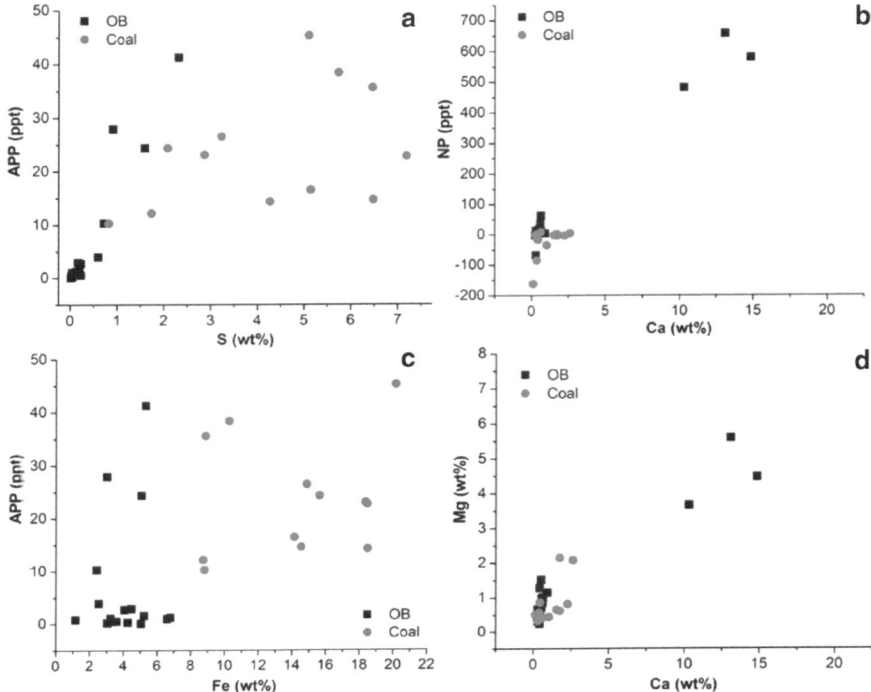

Fig. 11.3 Relationship between S and APP (**A**), NP and Ca (**B**), APP and Fe (**C**), and Mg and Ca (**D**) ir overburden and coal from Makum coalfield

The NNP and NPR are commonly used to assess the post-mining water quality more accurately than both NP and APP [5]. The criteria for evaluating mine discharge quality based on NNP and NPR are given in Table 11.4. Based on these, all the coal measures are found to have potential for acid generating (Fig. 11.4). However, based on NPR value, it is found that most of the overburdens can generate net alkalinity except two samples, R-5 and R-9, showing negative NPR. The NNP of most of overburden ranged between 0 and 12 ppt indicating their capacity for generating either acidic or alkaline discharges; but in case of few samples it is much above the threshold for generation of alkaline discharge. NNP result is also negative for the same overburden samples which have acid generating capacity by using NPR. However, the NPR shows a good relationship with paste pH. Therefore, the NPR can be used as better predictor for the post mining discharge quality than NNP.

Table 11.4 Criteria for the characterization of post mining water quality

Water Quality	[a]NNP (ppt)	[b]NPR
Net Acid	<0	<1
Either acid, neutral or alkaline	0–12	1–2
Net Alkaline	>12	>2

[a]Brady et al. [28], [b]Perry [31]

Fig. 11.4 Relationship between paste pH and NNP (**A**); paste pH and NPR (**B**)

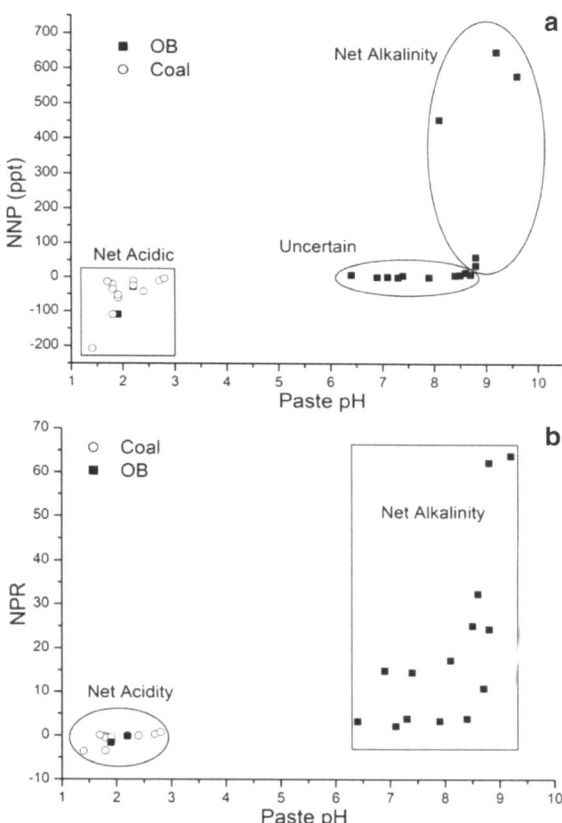

11.4.3   *Metal Concentration and Leaching Study*

Among the heavy metals in overburden, it is found that Mn has the highest concentration range, 158–1623 mg/kg, followed by Cr (105–433 mg/kg), Ni (41–309 mg/kg), Zn (5.8–199 mg/kg), Cu (4.5–66 mg/kg) and Pb (13–38 mg/kg). Most of these metals are above their respective crustal abundances and Ni showed the highest enrichment followed by Cr (Fig. 11.5). Similarly, in coal Cr ranges from 0.2 to 26.6 mg/kg, Cu 3.1–48.5 mg/kg, Mn 7.8–115.9 mg/kg, Ni 2.5–211 mg/kg, Pb 14.6–36.9 mg/kg and

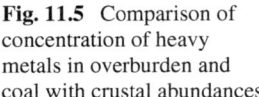

Fig. 11.5 Comparison of concentration of heavy metals in overburden and coal with crustal abundances

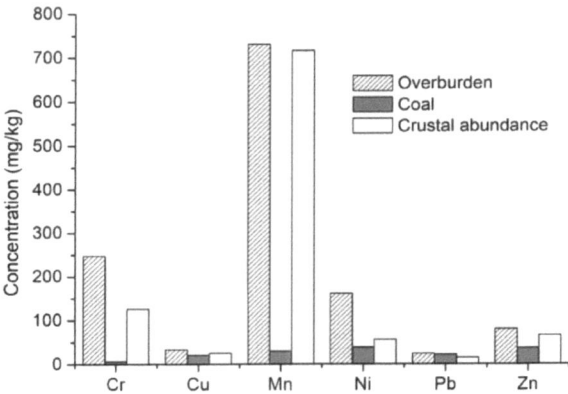

Zr 11.3–99.4 mg/kg. In contrast to that of overburden, except Pb, all other heavy metals are below their respective crustal abundances in coal.

Time-dependent leaching of metals from overburden and coal was carried out as described earlier. The change in pH and concentrations of metals leached with time are given in Table 11.5. During the leaching of sandstone and siderite-bearing shale, the initial pH of leachate is found to be acidic; however, the pH increases with time and becomes alkaline after 6 days (Fig. 11.6). It is due to dissolution of carbonates present in trace amounts and consumption of H⁺ by silicates. Shale with higher dolomite content has produced alkaline leachate. After the day 1 the leachate pH has been found to be 7.6 and increased to 8.1 at the end of 9 days. Presence of higher amount of dolomite in overburden has consumed the initial acidity and leachate becomes alkaline. Coal and overburden, those are rich with pyrite, have generated acidic leachate throughout the experiments. The leachate pH has been found in the range of 2.1–2.5 and 1.6–2.9 for coal and pyritic-shale respectively. The leachate from pyrite-rich overburden and coal remains acidic due to the oxidation of pyrite that is responsible to produce acidity (Fig. 11.6).

The concentrations of Mn, Ni and Pb in the leachate has been observed to be higher than their respective water quality guideline values as per the Bureau of India Standards (BIS) (Table 11.6), while concentrations of Cr, Cu and Zn are within their respective limits in leachate from both overburden and coal. Some overburden releases Cr above its permissible limit. Similar to the result obtained from leaching experiments, it has been commonly observed that Mn, Ni and Pb exceed their permissible limits in different coal mine drainages from USA and other countries [17, 32–34]. Equeenuddin et al. [17] reported the high concentration of Mn, Ni and Pb in the mine discharges and ground water in the Makum coalfield region.

The leachate generated from the coal and pyrite-rich overburden contains abnormally high metal concentrations with respect to that of the other overburdens. It is attributed to the oxidation of pyrite that causes very low pH (<3). Thus, metals are highly mobilized under strongly acidic environment as solubility of metals increases with decrease in pH [35]. The concentration of Mn in the leachate obtained from

Table 11.5 pH and concentration of metals (mg/L) in leachate at different time intervals using acidify deionized water at pH 4.2

Type	Days	1	2	4	6	9
R1 (Siderite bearing shale)	pH	4.60	4.70	6.90	7.00	7.80
	Zn	0.15	0.10	0.06	0.03	0.03
	Cu	ND	ND	ND	ND	ND
	Mn	1.45	1.76	1.76	1.37	1.37
	Ni	0.39	0.19	0.12	0.03	0.03
	Cr	0.08	0.08	0.07	0.06	0.06
	Pb	0.19	0.13	0.11	0.18	0.24
	Days	1	2	4	6	9
R5 (Pyrite bearing shale)	pH	2.90	2.50	3.00	1.60	2.90
	Zn	3.85	2.91	2.70	2.95	2.87
	Cu	1.48	0.99	1.12	1.20	1.19
	Mn	5.27	4.47	4.25	5.04	4.89
	Ni	4.95	3.88	3.68	3.85	3.67
	Cr	0.28	0.28	0.28	0.32	0.29
	Pb	0.30	0.24	0.16	0.27	0.27
	Days	1	2	4	6	9
R12 (Dolomite bearing shale)	pH	7.60	7.40	8.00	8.80	8.10
	Zn	ND	0.05	0.04	0.02	0.01
	Cu	ND	ND	ND	ND	ND
	Mn	0.32	0.30	0.32	0.23	0.21
	Ni	0.09	0.09	0.06	0.15	0.02
	Cr	0.05	0.06	0.05	0.04	0.04
	Pb	0.15	0.15	0.08	0.19	0.21
	Days	1	2	4	6	9
R16 (Sandstone)	pH	4.40	4.90	6.50	6.80	7.30
	Zn	0.02	0.01	0.01	0.01	0.01
	Cu	ND	ND	ND	ND	ND
	Mn	0.14	0.18	0.15	0.13	0.12
	Ni	0.12	0.13	0.11	0.11	0.13
	Cr	0.03	0.04	0.04	0.03	0.03
	Pb	0.09	0.07	0.06	0.12	0.14
	Days	1	2	4	6	9
Coal	pH	2.50	2.50	2.40	2.30	2.10
	Zn	1.93	1.81	1.86	1.99	1.95
	Cu	0.38	0.38	0.36	0.44	0.48
	Mn	2.84	2.63	2.78	3.10	2.91

(continued)

Table 11.5 (continued)

Type	Days	1	2	4	6	9
	Ni	13.95	13.39	13.65	14.45	13.76
	Cr	BDL	BDL	BDL	BDL	BDL
	Pb	0.23	0.21	0.26	0.28	0.24

Fig. 11.6 Variation of leachate pH with time for different overburden and coal

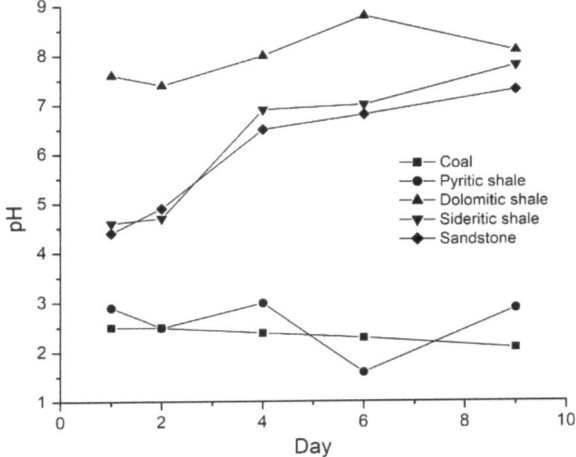

Table 11.6 Desirable limits of drinking water quality as per Indian Standard (IS: 10500; BIS 2012)

Parameters	Limits
pH	6.5–8.5
Zn	15
Cu	1.5
Mn	0.3
Ni	0.02
Cr	0.05
Pb	0.01

siderite dominated overburden is very high (up to 1.76 mg/L) as compared to other non-pyritic ones (up to 0.32 mg/L). It is caused due to its possible association with siderite. Occurrence of Mn in siderite bearing overburden has been earlier reported and indicated siderite as the source for very high concentration of Mn in coal mine drainage [36]. Similarly, abnormally high concentration of Ni (up to 14.5 mg/L) has been found in the coal leachate compared to that of the overburden (up to 4.9 mg/L). This might be due the occurrence of Ni in the exchangebale part of the coal components as it is easily water soluble. Finkelman et al. [37] reported the association of 55% of the Ni in an exchangeable form in coal. Equeenuddin et al. [17]

reported abnormally high concentration of Ni (up to 11.13 mg/L) in Makum coal mine drainages relative to that of other places. Therefore, it can be concluded that the results of the leachate chemistry from the mine overburden and coal provides very good information on mine dishcharges and the likely potential metal contamination due to coal mine drainages.

11.5 Conclusion

This study aims to identify the mineralogy of overburden and coal in order to evaluate the acid mine drainage potential vis-á-vis metal leaching characteristics for understanding the impact of mining activities in water, soil and sediment in the Makum coalfield. Dolomite is found to be abundant in the overburden and the most important mineral contributing towards the maximum NP. The coal measures from all collieries are highly acid producing. Siltstones and massive carbonaceous shale are observed to have higher neutralization potential whereas laminated carbonaceous shale are found to be enriched in pyrite and have maximum APP. The results of ABA test is nearly consistent with the direct mine discharge and NPR is found to predict the mine discharge water quality more accurately that that of NNP. Leaching study indicates that the concentration of Mn, Ni and Pb is above their respective allowable limits. Thus, it is proposed that ABA study along with the leaching test of overburden and coal is highly essential to assess the post-mining water quality and its environmental impact.

References

1. Gray, N. F. (1997). Environmental impact and remediation of acid mine drainage: A management problem. *Environmental Geology, 30*, 62–71.
2. Guseva, O., Opitz, A. K., Broadhurst, J. L., Harrison, S. T., & Becker, M. (2021). Characterisation and prediction of acid rock drainage potential in waste rock: Value of integrating quantitative mineralogical and textural measurements. *Minerals Engineering, 163*, 106750.
3. Jambor, J. L., Dutrizac, J. E., Raudsepp, M., & Groat, L. A. (2003). Effect of peroxide on neutralization potential values of siderite and other carbonate minerals. *Journal of Environmental Quality, 32*(6), 2373–2378.
4. Perry, E. (1985). Overburden analysis: An evaluation methods. In *Proceedings Symposium of Surface Mining, Hydrology, Sedimentology and Reclamation* (pp. 369–375).
5. Skousen, J., Simmons, J., McDonald, L. M., & Ziemkeiwicz, P. (2002). Acid-base accounting to predict post-mining drainage quality on surface mines. *Journal of Environmental Quality, 31*, 2034–2044.
6. Sherlock, E. J., Lawrence, R. W., & Poulin, R. (1995). On the neutralization of acid rock drainage by carbonate and silicate minerals. *Environmental Geology, 25*, 43–54.
7. Perry, E. F., & Brady, K. B. C. (1995). Influence of neutralization potential on surface mine drainage in Pennsylvania. In *Proceedings Sixteenth Annual West Virginia Surface Mine Drainage Task Force Symposium* (p. 16).

8. Jambor, J. L., Dutrizac, J. E., Groat, L. A., & Raudsepp, M. (2002). Static tests of neutralization potentials of silicate and aluminosilicate minerals. *Environmental Geology, 43*, 1–17.

9. Jambor, J. L., Dutrizac, J. E., & Raudsepp, M. (2007). Measured and computed neutralization potentials from static tests of diverse rock types. *Environmental Geology, 52*, 1019–1031.

10. Skousen, J., Renton, J., Brown, H., Evans, P., Leavitt, B., Brady, K., Cohen, L., & Ziemkiewicz, P. (1997). Neutralization potential of overburden samples containing siderite. *Journal of Environmental Quality, 26*, 673–681.

11. Daniels, W. L., Zipper, C. E., & Orndorff, Z. W. (2014). Predicting release and aquatic effects of total dissolved solids from Appalachian USA coal mines. *International Journal of Coal Science Technology, 1*(2), 152–162.

12. Shan, Y., Wang, W., Qin, Y., & Gao, L. (2019). Multivariate analysis of trace elements leaching from coal and host rock. *Groundwater for Sustainable Development, 8*, 402–412.

13. Equeenuddin, S. M., Tripathy, S., Sahoo, P. K., & Panigrahi, M. K. (2013). Metal behavior in sediment associated with acid mine drainage stream: Role of pH. *Journal of Geochemical Exploration, 124*, 230–237.

14. Sahoo, P. K., Equeenuddin, S. K., & Powell, M. A. (2016). Trace elements in soils around coal mines: Current scenario, impact and available techniques for management. *Current Pollution Reports, 2*, 1–14.

15. Zhou, C., Liu, G., Wu, D., Fang, T., Wang, R., & Fan, X. (2014). Mobility behavior and environmental implications of trace elements associated with coal gangue: A case study at the Huainan Coalfield in China. *Chemosphere, 95*, 193–199.

16. Quevauviller, P. H., Van Der Sloot, H. A., Ure, A., Muntau, H., Gomez, A., & Rauret, G. (1996). Conclusions of the workshop: Harmonization of leaching/extraction tests for environmental risk assessment. *Science of the Total Environment, 178*(1–3), 133–139.

17. Equeenuddin, S. M., Tripathy, S., Sahoo, P. K., & Panigrahi, M. K. (2010). Hydrogeochemical characteristics of acid mine drainage and water pollution at Makum coalfield, India. *Journal of Geochemical Exploration, 105*(3), 75–82.

18. Rawat, N. S., & Singh, G. (1982). The role of micro-organism in the formation of acid mine drainage in the north eastern coal field of India. *International Journal of Mine Water, 1*(2), 29–36.

19. Ahmed, M. (1996). Petrology of Oligocene coal, Makum coalfield, Assam, northeast India. *International Journal of Coal Geology, 30*(4), 319–325.

20. Baruah, B. P., Saikia, B. K., Kotoky, P., & Rao, P. G. (2006). Aqueous leaching of high sulfur sub-bituminous coals in Assam, India. *Energy & Fuels, 20*(4), 1550–1555.

21. Baruah, B. P., Kotoky, P., & Borah, G. C. (2003). Distribution and nature of organic/mineral bound elements in Assam coals, India. *Fuel, 82*(14), 1783–1791.

22. Barooah, P. K., & Baruah, M. K. (1996). Sulphur in Assam coal. *Fuel Processing Technology, 46*(2), 83–97.

23. Rajarathnam, S., Chandra, D., & Handique, G. K. (1996). An overview of chemical properties of marine-influenced Oligocene coal from the northeastern part of the Assam-Arakan basin, India. *International Journal of Coal Geology, 29*(4), 337–361.

24. Mukherjee, S., & Borthakur, P. C. (2003). Effect of leaching high sulphur subbituminous coal by potassium hydroxide and acid on removal of mineral matter and sulphur. *Fuel, 82*(7), 783–788.

25. Chandra, D., Chaudhuri, S. G., & Ghose, S. (1980). Distribution of sulphur in coal seams with special reference to the Tertiary coals of North-Eastern India. *Fuel, 59*(5), 357–359.

26. Price, W. A., Morin, K., Hutt, N. (1997). Guidelines for the prediction of acid-rock drainage and metal leaching for mines in British Columbia: Part II recommended procedures for static and kinetic testing. In *Proceedings of the 4th International Conference on Acid Rock Drainage* (pp. 15–30).

27. Sobek, A., Schuller, W., Freeman, J. R., & Smith, R. M. (1978). *Field and laboratory methods applicable to overburdens and minesoils.* Industrial Environmental Research Laboratory, Office of Research and Development, US Environmental Protection Agency Report, EPA-600/2-78-054, Cincinnati, Ohio.

28. Brady, K. B. C., Perry, E. F., Beam, R. L., Bisko, D. C., Gardner, M. D., Tarantino, J. M. (1994). Evaluation of acid base accounting to predict the quality of drainage at surface coal mines in Pennsylvania, U.S.A. In *International Land Reclamation and Mine Drainage Conference on the Abatement of Acidic Drainage. Vol. 1: Mine drainage—SP 06A-94* (pp. 138–147).
29. Sobek, A. A., Skousen, J. G., & Scott, E. F. (2000). Chemical and physical properties of overburdens and minesoils. In R. I. Barnhisel, R. G. Darmody, & W. L. Daniels (Eds.), *Reclamation of drastically disturbed lands. Agronomy monograph number 41* (pp. 77–104). American Society of Agronomy, Crop Science Society of America, & Soil Science Society of America.
30. Haney, E. B., Haney, R. L., Hossner, L. R., & White, G. N. (2006). Neutralization potential determination of siderite ($FeCO_3$) using selective oxidants. *Journal of Environmental Quality, 35*(3), 871–879.
31. Perry, E. (1998). Interpretation of acid-base accounting. Chapter 11. In *Coal mine drainage prediction and pollution prevention in Pennsylvania*. Pennsylvania Department of Environment Protection.
32. Chon, H. T., Kim, J. Y., & Choi, S. Y. (1999). Hydrogeochemical characteristics of acid mine drainage around the abandoned Youngdong Coal Mine in Korea. *Resource Geology, 49*(2), 113–120.
33. Evangelou, V. P. (1998). *Environmental soil and water chemistry: Principles and applications.* Wiley & Sons.
34. Sahoo, P. K., Tripathy, S., Equeenuddin, S. M., & Panigrahi, M. K. (2012). Geochemical characteristics of coal mine discharge vis-à-vis behavior of rare earth elements at Jaintia Hills coalfield, northeastern India. *Journal of Geochemical Exploration, 112*, 235–243.
35. Saria, L., Shimaoka, T., & Miyawaki, K. (2006). Leaching of heavy metals in acid mine drainage. *Waste Management & Research, 24*(2), 134–140.
36. Larsen, D., & Mann, R. (2005) Origin of high manganese concentrations in coal mine drainage, eastern Tennessee. *Journal of Geochemical Exploration, 86*, 143–163.
37. Finkelman, R. B., Palmer, C. A., Krasnow, M. R., Aruscavage, P. J., Sellers, G. A., & Dulong, F. T. (1990). Combustion and leaching behavior of elements in the Argonne premium coal samples. *Energy & Fuels, 4*(6), 755–766.